飞行器设计与制造专业英语

Professional English in Aircraft Design and Manufacture

黄晓明 冯伟伟 编著

西北工业大学出版社
西 安

【内容简介】 本书主要介绍航空宇航科学与技术专业人员从事该领域工作的必备知识。本书内容包含飞行器结构、飞行原理、飞行器控制、飞行器先进制造技术、飞机装配、无人机以及CAD/CAM技术等专业实例,是一本全面介绍飞行器设计与制造专业英语的图书。

本书可作为高等院校飞行器设计、航空宇航制造工程等多个航空宇航科学与技术类本科以及研究生的教学参考书,也可供有关专业研究人员及生产人员参考应用。

图书在版编目(CIP)数据

飞行器设计与制造专业英语/黄晓明,冯伟伟著
.—西安:西北工业大学出版社,2018.10
ISBN 978-7-5612-6309-9

Ⅰ.①飞… Ⅱ.①黄… ②冯… Ⅲ.①飞行器-英语
Ⅳ.①V47

中国版本图书馆CIP数据核字(2018)第238113号

飞 行 器 设 计 与 制 造 专 业 英 语
FEIXINGQI SHEJI YU ZHIZAO ZHUANYE YINGYU

策划编辑:孙显章
责任编辑:王梦妮 何格夫

出版发行:西北工业大学出版社
通信地址:西安市友谊西路127号 邮编:710072
电 话:(029)88493844 88491757
网 址:www.nwpup.com
印 刷 者:陕西金德佳印务有限公司
开 本:787 mm×1 092 mm 1/16
印 张:8
字 数:134千字
版 次:2018年10月第1版 2018年10月第1次印刷
定 价:30.00元

前　　言

　　《飞行器设计与制造专业英语》主要以英语课文的形式介绍飞行器设计与制造领域的基本概念、原理和相关知识，以培养飞行器设计制造工程专业人员的专业英语水平为主要目标，使其掌握一般飞行器设计及制造所需的常用英语，借助于工具阅读或翻译相关专业文献。

　　飞行器设计与制造技术发展迅猛，其包含的学科越来越庞杂，知识更新速度也越来越快。因此，本书按照实用、新颖、系统的原则，精选具有专业特点和代表性的专业内容。全书共设定10个单元模块，各部分相对独立，包括飞行器结构、飞行原理、飞行器控制、飞行器材料、飞行器先进制造技术、无人机以及科技文献写作等。每个单元包括阅读材料、专业词汇、注释及参考译文。本书可作为高等院校飞行器设计制造类专业英语教材或课外阅读材料，也可供航空飞行器广大专业技术人员学习参考。需要说明的是，本书中给出的译文并不是唯一的，仅作为参考。

　　本书的编著过程中，得到了滨州学院航空工程一流学科的资助，在此表示感谢。

　　由于水平限制，书中缺点在所难免，恳请批评指正。

编　者

2018年4月

目　　录

目录

Unit One

Passage One: Aircraft Structures and Materials

1. How does an aircraft balance weight and strength?

Finding a safe compromise between low weight and high strength is critical when creating an aircraft. Aircraft structures must be light yet strong and stiff enough to resist the various forces acting on an airplane during flight. They must also be durable enough to withstand these forces over the airplane's entire life span.

A big focus in the design of airplanes is to make them weigh as little as possible. Materials engineers study materials, both conventional and composite for use in airplane structures. Some areas of concern are the strength and rigidity of the material, its availability, its ease of processing, and its resistance to temperature and fatigue.

Any aircraft design is a carefully planned compromise in which many competing factors are traded against one another: payload capacity, cost, range, speed, fuel economy, durability, noise levels, required runway length, and many others. The function of an aircraft, whether an airliner or a fighter, a business jet or a private airplane, is the major influence in balancing these factors. The best design typically provides maximum performance at the lowest weight.

2. What kinds of materials are used to make aircraft?

Several materials can be used for major structures—wings, fuselage, or landing gear for example—on different types of aircraft.

Wood was used on most early airplanes and is now mainly used on homebuilt airplanes. Wood is lightweight and strong, but it also splinters and requires a lot of maintenance.

Aluminum is used on most types of aircraft because it is lightweight and strong. Aluminum alloys don't corrode as readily as steel. But because they lose their strength at high temperatures, they cannot be used for skin surfaces that become very hot on airplanes that fly faster than twice the speed of sound.

Steel can be up to four times stronger and three times stiffer than aluminum, but it is also three times heavier. It is used for certain components like landing gear, where strength and hardness are especially important. It has also been used for the skin of some high-speed airplanes, because it holds its strength at higher temperatures better than aluminum.

Graphite-epoxy is one of several types of composite materials that are becoming widely used for many aircraft structures and components. These materials typically consist of strong fibers embedded in a resin (in this case, graphite fibers embedded in epoxy). Thin sheets of the material can be stacked in various ways to meet specific strength or stiffness needs. Graphite-epoxy is about as strong as aluminum and weighs about half as much.

Titanium is about as strong as steel and weighs less, though it is not as light as aluminum. It holds its strength at high temperatures and resists corrosion better than steel or aluminum. Though titanium is expensive, these characteristics have led to its greater use in modern aircraft.

After repeated bending and twisting, a small crack can form and grow where stress is concentrated. That's why aircraft manufacturers use nondestructive evaluation (NDE) techniques while making an aircraft, and throughout an aircraft's life, to detect cracks. NDE techniques are used to evaluate the properties of materials without causing damage. A commonly used method of NDE is ultrasound scanning, which analyzes echoes from ultrasonic waves to reveal problems inside the material.

3. What happened to vehicles at hypersonic speeds?

A vehicle designed to fly at hypersonic speed (more than five times the speed of sound) must be able to withstand searing heat due to friction and shock waves. Special materials must be used to protect it and insulate the people and instruments on board from the heat.

What happens to air at high speeds?A fast-moving vehicle compresses the air in front of it, causing the air's temperature to rise.

Combined with friction, this heat becomes tremendous at extremely high speeds. The Apollo 11 spacecraft reentered the atmosphere traveling about 11 kilometers (7 miles) per second. The air just in front of it made a fireball hotter than the surface of the Sun!

Insulating materials, like the tiles on the Space Shuttle, shield a spacecraft from the heat. A spacecraft's shape also plays a role. Besides slowing the vehicle by creating drag, a blunt shape helps to maintain an insulating cushion of air between the spacecraft and the shock wave it creates.

Another way to protect a vehicle from reentry heating is by using an ablation shield, like those on the Mercury, Gemini and Apollo spacecraft. These shields were covered with a material that burned off (ablated) at extremely high temperatures and dissipated the heat. Unlike the tiles on the Shuttle, ablation shields can only be used once.

Space Shuttle tiles are made of a lightweight, ceramic material. They are used on the Shuttle to insulate the vehicle from the searing heat of reentry.

Sufficient stress will buckle—and in some cases can destroy—a structure. Aircraft are designed to resist such buckling. Buckling depends not only on the physical properties of the structural material but also on thickness and shape.

If a spacecraft flies only in space and not through an atmosphere, it does not have to deal with drag. Keeping weight down is important, because of the high energy and cost needed to send even small amounts of mass into space. If manned, a spacecraft must have enough shielding to protect its occupants from such hazards as solar radiation.

参考译文：飞行器结构与材料

1. 飞机如何平衡重量和强度之间的关系？

寻找低重量和高强度之间的安全折中点是制造飞机的关键。飞机结构必须轻巧，又要足够坚固，以抵抗在飞行中作用在飞机上的各种力。它们还必须具有足够的耐用性，能够承受飞机整个寿命期间的这些力。

飞机设计的重点在于尽可能减轻飞机的重量。材料工程师研究用于飞机结构设计的传统及复合材料。关注的领域是材料的强度和刚度，其可用性，易于加工以及耐温度和疲劳性等方面。

任何飞机设计方案都是经过精心策划的折中方案，其中许多因素需要相互协调以达平衡，比如有效载荷能力、成本、航程、速度、燃油经济性、耐久性、噪声水平、所需的跑道长度等。飞机的功能，无论是客机还是战机、商用喷气式飞机或私人飞机，都是平衡这些因素的主要方面。最佳设计通常表现为重量最轻而性能最高。

2. 用什么材料制造飞机？

可以将几种不同类型的材料用于飞机的主要结构，例如，机翼、机身或起落架。

早期的大多数飞机都使用木材，现在木材则主要用于自制飞机上。木材质轻，坚固，但也容易分裂，需要长期维护。

大多数类型的飞机使用铝，因为其重量轻，强度高。铝合金不像钢一样易腐蚀。但是因为它们在高温下易失去强度，所以不能用于飞行速度超过声速两倍的飞机的外壳。

钢的强度是铝的四倍，硬度是铝的三倍，但是重量是铝的三倍。它用于某些部件，如起落架，该部位强度和硬度尤其重要。它也被用于一些高速飞机的外壳，因为在高温下与铝相比，它保持强度的性能更强。

石墨-环氧树脂是几种复合材料之一，其被广泛用于许多飞机结构和部件。这些材料通常由嵌入树脂中的强纤维（在这种情况下，是嵌入环氧树脂中的石墨纤维）组成。薄片型的材料可以以各种方式堆叠以满足特定的强度或刚度需求。石墨-环氧树脂与铝强度相当，重量约为铝的一半。

钛的强度与钢一样，但重量更小，尽管不如铝那么轻。却能在高温下保持强度，比钢或铝更耐腐蚀。虽然钛很昂贵，但它的这些特征使之在现代飞机中得到了更多的应用。

经过反复弯曲和扭折，应力集中处可萌生小裂纹。这就是飞机制造商在制造飞机并且在整个飞机寿命中使用非破坏性评估（NDE）技术来检测裂缝的原因。NDE技术用于评估材料的性质而不会对其造成损坏。NDE的常用方法是超声波扫描，它能分析来自超声波的回波以揭示材料内部存在的问题。

3. 高超音速飞行器如何飞行？

设计的高超音速飞行器（超过声速的五倍）必须能够承受大气环境中由于摩擦和冲击波而产生的热量，而且必须使用特殊材料进行保护，并将人体和仪器与热量隔绝。

空气在高速环境中如何变化？快速移动的飞行器将其前面的空气压缩，导致空气的温度升高。

由摩擦引起的热量，在非常高的速度下变得巨大。阿波罗11宇宙飞船重新进入大气中时每秒行驶约11千米（7英里）。它前面的空气形成的火球比太阳的表面更热。

绝缘材料，如航天飞机上的瓷砖，可以保护航天器远离高温。航天器的形状也能发挥作用。除了通过制造向后的力使飞机减速，钝的形状有助于在航天器与其产生的冲击波之间形成并保持一个绝缘的空气垫。

保护飞行器免受折返热的另一种方法是使用消融护罩，如水星、双子座和阿波罗航天器上的那些消融护罩。这些护罩表层材料在极高温度下烧掉（烧蚀），并散发热量。不同于航天器上的瓷砖，消融护罩只能使用一次。

航天飞机隔热瓦由轻质陶瓷材料制成。它们用于航天飞机上，以隔绝飞机与飞行中的大气热量。

过大的压力会造成飞机结构扭曲，在某些情况下会破坏结构。飞机的设计需要能抵抗这种弯曲。是否易弯曲不仅取决于结构材料的物理性能，还取决于它的厚度和形状。

如果航天器只在空间飞行，而不是通过大气层，则不必进行处理。减轻航天器重量是很重要的，因为将较轻质量的物体送入太空所需的能量和成本低。如果载人，航天器必须设置足够的屏蔽装置，以保护航天员免受太阳辐射等危害。

New Words and Expressions:

compromise ['ka:prəmaɪz]vt.妥协；危害

rigidity[rɪ'dʒɪdətɪ]n. [物] 硬度；[力] 刚性；严格，刻板；僵化；坚硬

resistance[rɪ'zɪst(ə)ns]n.阻力；电阻；抵抗；反抗；抵抗力

payload capacity 额定载重量；有效负载能力

durability['djʊrə'bɪləti]n. 耐久性；坚固；耐用年限

maximum ['mæksɪməm]n. [数] 极大，最大限度；最大量

fuselage ['fju:zəlɑ:ʒ; -lɪdʒ]*n.*[航] 机身（飞机）

landing gear[航] 起落架；起落装置，着陆装置

splinters ['splɪntə]*n.* 碎片；微小的东西；极瘦的人

aluminum [ˌæljʊ'mɪnɪəm]*n.* 铝

alloy ['ælɔɪ] *vt.*使成合金；使减低成色

corrode[kə'rəʊd] *vt.*侵蚀；损害

Graphite-epoxy 石墨纤维树脂

fiber ['faɪbər] *n.*纤维；光纤

graphite ['græfaɪt]*n.*石墨；黑铅

epoxy[ɪ'pa:ksɪ; e-]*adj.*环氧的

titanium[taɪ'teɪnɪəm; tɪ-]*n.* [化学] 钛

ultrasonic [ˌʌltrə'sa:nɪk] *adj.*[声] 超声的；超音速的

hypersonic [haɪpə'sa:nɪk] *adj.*极超音速的；高超音速动；[声] 特超声速的

shuttle['ʃʌt(ə)l] *n.* 航天飞机；穿梭；梭子；穿梭班机；公共汽车等

Questions:

Answer the following questions according to the text.

（1）How does an aircraft balance weight and strength?

（2）How do astronauts survive the heat of reenter?

（3）What happened to vehicles at hypersonic speeds?

Passage Two:The Requirements of Modern Aircraft Design

Modern aircraft design focuses on the integration of new technologies and systems with current and advanced configurations. This includes new structures, materials and manufacturing processes. The goal is to move towards environmentally—friendly and cost—effective aviation in the civil arena and high-performance and effective aviation in the military arena.

New aircraft design is essential to address issues such as carbon footprint reduction, lower noise pollution and improved passenger comfort; as well as contributing to national security.

Aircraft design is an essential element in the effort to reduce aviation's environmental impact. Researchers, designers and manufacturers are advancing aerodynamics, structures and materials, control systems and propulsion in order to make planes cleaner, quieter and more efficient. In the last 50 years, these advancements have been successfully achieved predominantly by evolutionary improvements in aircraft designs, but there are also solutions around which would radically change aircraft configurations .

In addition to the options mentioned above, configuring aircraft and engines that support the use of alternative fuels has the potential to reduce greenhouse gas and air quality-related emissions while diversifying energy supplies. However, some of these fuel options such as liquid natural gas (LNG) require modifications in terms of aircraft and engine design and, therefore, come at an additional cost.

The Laboratory for Aviation and the Environment is currently investigating the technical, environmental and economic challenges for introducing a fleet of aircraft that burn LNG as a second fuel. It has developed system-level solutions on how aircraft could be retrofitted to accomodate dual fuel-use. The laboratory has advanced computational tools to evaluate the performance of dual-fuel use aircraft and to assess the full environmental and ecomomic impact of deploying these aircraft.

Solutions on how to accomodate LNG as a supplemental fuel in existing aircraft are still in their early stages and the feasibility and viability of dual fuel-use needs to thoroughly assessed. Research gives important advice to aircraft manufactures and operators, and may lead to a new type of aircraft entering the fleet – with long-lasting implications for aviation markets in general.

参考译文：现代飞机设计要求

现代飞机设计侧重于新技术和现代高级配置系统的结合。这包括新的结构，材料和制造工艺。现代飞机设计的目标便是使其在民航领域的发展更加环保和低成本，在军事领域的发展具有更高的性能和效率。

新的飞机设计对于解决诸如减少碳排放，降低噪声污染和改善乘客舒适度等问题至关重要，并能够为国家安全做出贡献。

飞机设计对于减少对航空环境的影响至关重要。研究人员、设计师和制造商致力于推进空气动力学、结构和材料、控制系统、推进力系统等方面研究，以使飞机更清洁、更安静、更高效。在过去50年中，这些进步主要是通过飞机设计的改进而得以成功实现的，但该方面的有些解决方案将会彻底改变飞机配置。

除了上述选项之外，某些支持使用替代燃料的飞机和发动机有可能在影响能源供应多样化的同时减少温室气体以及影响空气质量的气体的排放。然而，这其中的某些燃料，例如液化天然气（LNG）的使用则需要对飞机和发动机设计进行修改，因此需要额外的成本。

航空与环境实验室目前正在调查引进LNG作为第二种燃料的飞机所面临的技术、环境和经济等方面的挑战。它已经开发了关于如何改装飞机以适应双重燃料使用的系统级解决方案。该实验室拥有先进的计算工具且能评估双燃料使用型飞机的性能，并评估这些部署对环境和经济的全面影响。

关于如何将液化天然气作为现有飞机补充燃料的解决方案的有关研究目前仍处于早期阶段，双燃料使用的可行性还需要进行彻底的评估。这些研究为飞机制造商和运营商提供了重要的依据，可能导致一种新型飞机进入市场，并会对航空市场产生持续影响。

New Words and Expressions:

integration [ˌɪntɪ'greɪʃ(ə)n]*n.*集成；综合

technology [tek'na:lədʒɪ]*n.* 技术；工艺；术语

configuration [kənˌfɪgə'reɪʃ(ə)n; -gjʊ-]*n.*配置；结构；外形

manufacturing[ˌmænjʊ'fæktʃərɪŋ]*adj.*制造业；制造业的

environmentally-friendly *adj.* 对环境无害的；保护生态环境

cost-effective['kɔ:stə'fektɪv]*adj.*划算的；成本效益好的

arena [ə'ri:nə]*n.*舞台；竞技场

carbon footprint 碳足迹；碳排放量

element ['elɪm(ə)nt]*n.*元素；要素；原理；成分；自然环境

predominantly [prɪ'da:mɪnəntlɪ]*adv.* 主要地；显著地

greenhouse gas *n.*二氧化碳、甲烷等导致温室效应的气体

diversify [daɪ'vɜːsɪˌfaɪ]*vt.* 使多样化，使变化；增加产品种类以扩大

liquid natural gas 液化天然气

hybrid['haɪbrɪd]*n.* 杂种，混血儿；混合物

system-level 系统级

retrofit ['retrəʊfɪt] *vt.*改进；[计] 更新；式样翻新

dual-fuel 双燃料

Questions:

Answer the following questions according to the text.

(1) What is the focus of modern aircraft design?

(2) What is the impact of modern aircraft design on the environment?

(3) What is a dual fuel aircraft?

Unit Two

Passage One: The Principle of Flight (1)

Almost everyone today has flown in an airplane. Many ask the simple question, "what makes an airplane fly?" The answer one frequently gets is misleading and often just plain wrong. We are going to show you that lift is easier to understand if one starts with Newton rather than Bernoulli. We will also show you that the popular explanation that most of us were taught is misleading at best and that lift is due to the wing diverting air down.

The first we will call the Mathematical Aerodynamics Description which is used by aeronautical engineers. This description uses complex mathematics and/or computer simulations to calculate the lift of a wing. These are design tools which are powerful for computing lift but do not lend themselves to an intuitive understanding of flight.

The second description we will call the Popular Explanation which is based on the Bernoulli principle. The primary advantage of this description is that it is easy to understand and has been taught for many years. Because of its simplicity, it is used to describe lift in most flight training manuals. The major disadvantage is that it relies on the "principle of equal transit times" which is wrong. This description focuses on the shape of the wing and prevents one from understanding such important phenomena as inverted flight, power, ground effect, and the dependence of lift on the angle of attack of the wing.

The third description, which we are advocating here, we will call the Physical

Description of lift. This description is based primarily on Newton's laws. The physical description is useful for understanding flight. Little math is needed to yield an estimate of many phenomena associated with flight. This description gives a clear, intuitive understanding of such phenomena as the power curve, ground effect and high-speed stalls. However, unlike the mathematical aerodynamics description, the physical description has no design or simulation capabilities.

1. The popular explanation of lift

Students of physics and aerodynamics are taught that airplanes fly as a result of Bernoulli's principle, which says that if air speeds up the pressure is lowered. Thus a wing generates lift because the air goes faster over the top creating a region of low pressure, and thus lift. This explanation usually satisfies the curious and few challenge the conclusions. Some may wonder why the air goes faster over the top of the wing.

In order to explain why the air goes faster over the top of the wing, many have resorted to the geometric argument that the distance the air must travel is directly related to its speed. The usual claim is that when the air separates at the leading edge, the part that goes over the top must converge at the trailing edge with the part that goes under the bottom. This is the so-called "principle of equal transit times".

As discussed by Gale Craig, let us assume that this argument were true. The average speeds of the air over and under the wing are easily determined because we can measure the distances and thus the speeds can be calculated. From Bernoulli's principle, we can then determine the pressure forces and thus lift. If we do a simple calculation we would find that in order to generate the required lift for a typical small airplane, the distance over the top of the wing must be about 50% longer than under the bottom. Fig.2-1 shows what such an airfoil would look like.

Fig.2-1　Shape of wing predicted by principle of equal transit time

If we look at the wing of a typical small plane, which has a top surface that is 1.5%–2.5% longer than the bottom, we discover that a Cessna 172 would have to fly at over 400 mph to generate enough lift. Clearly, something in this description of lift is flawed.

But, who says the separated air must meet at the trailing edge at the same time? Fig.2-2 shows the airflow over a wing in a simulated wind tunnel. In the simulation, colored smoke is introduced periodically. One can see that the air that goes over the top of the wing gets to the trailing edge considerably before the air that goes under the wing. In fact, close inspection shows that the air going under the wing is slowed down from the "free-stream" velocity of the air. So much for the principle of equal transit times.

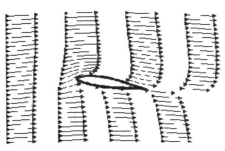

Fig.2-2　Simulation of the airflow over a wing in a wind tunnel, with colored "smoke" to show the acceleration and deceleration of the air

The popular explanation also implies that inverted flight is impossible. How a wing adjusts for the great changes in load such as when it is in a steep turn? So, why has the popular explanation prevailed for so long? One answer is that the Bernoulli principle is easy to understand. There is nothing wrong with the Bernoulli principle, or with the statement that the air goes faster over the top of the wing. But, as the above discussion suggests, our understanding is not complete with this explanation. The problem is that we are missing a vital piece when we apply Bernoulli's principle. We can calculate the pressures around the wing if we know the speed of the air over and under the wing.

2. Newton's laws and lift

So, how does a wing generate lift? To begin to understand lift we must return to high school physics and review Newton's first and third laws. Newton's first law states a body at rest will remain at rest, or a body in motion will continue in straight-line motion unless subjected to an external applied force. That means, if one sees a bend in the flow of air, or if air originally at rest is accelerated into motion, there is a force acting on it. Newton's third law states that for every action there is an equal and opposite reaction. As an example, an object sitting on a table exerts a force on the table and the table puts an equal and opposite force on the object to hold it up. In order to generate lift a wing must do something to the air.

Let's compare two figures used to show streams of air (streamlines) over a wing. In fig.2-3 the air comes straight at the wing, bends around it, and then leaves straight behind the wing. We have all seen similar pictures in flight manuals. But, the air leaves the wing exactly as it appeared ahead of the wing. There is no action on the air so there can be no lift! Fig.2-4 shows the streamlines, the air passes over the wing and is bent down. The bending of the air is the action of the lift on the wing.

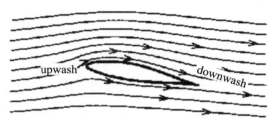

Fig.2-3   Common depiction of airflow over a wing.
This wing has no lift

Fig.2-4   True airflow over a wing with lift, showing
upwash and downwash

参考译文：飞行原理(1)

当今几乎每个人都坐过飞机。许多人会问一个简单的问题，"什么能使飞机飞行？"我们得到的答案往往是错误的或者是带有误导性的。我将要告诉你，如果是从牛顿定律而不是伯努利原理开始，那么飞行便更容易理解了。我们还会告诉你，其实我们大多数人被教导的机翼转动，使空气下压产生升力的流行解释是有误导性的。

第一个描述我们称之为航空工程师使用的数学空气动力学描述，该描述使用复杂的数学和/或计算机模拟来计算机翼的升力。这些设计工具虽然对于计算升力而言是强大的，但不能直观地了解飞行。

第二个描述我们称之为基于伯努利原理的流行说明。这个描述的主要优点是它很容易理解并已经被使用很多年。由于其简单性，便被大多数飞行训练手册用以描述飞行器的升降。主要的缺点则是其错误地依靠了"等效转化时间原则"。这种描述的重点是机翼的形状，它不利于人们了解倒转飞行、动力、地面效应以及机翼迎角对升力的影响等重要现象。

我们在这里提出的第三个描述，称之为升力的物理描述。这个描述主要基于牛顿定律。物理描述有助于理解飞行。这需要少许数学知识即可判断许多与飞行有关的现象。该描述给出了对功率曲线、地面效应和高速挡位等现象的清晰直观的理解。然而，与数学空气动力学描述不同的是，物理描述没有设计或模拟能力。

1. 升力的通用解释

学习有关物理学和空气动力学的学生一直被误导，飞机是由于伯努利的原理而飞行的，即空气加速压力降低。因为空气在顶部流通更快产生低压区域，从而使机翼产生升力，因此飞行器升高。这个解释能够满足不会对结论提出挑战的人的好奇心。可有些人可能会想，为什么空气流通在机翼的顶部变得更快。

为了解释为什么空气流通在机翼顶部变得更快，许多人采用几何论证，即空中行驶的距离与其速度直接相关。通常的要求是，当空气在前缘分离时，超过顶部的部分必须在后缘处会聚，而部分落在底部下方。这就是所谓的"等效转化时间原则"。

如Gale Craig所讨论的，我们假设这个论点是真实的。因为我们可以测量距离，从而可以计算出机翼上下空气的平均速度。根据伯努利原理，我们可以确定压力，从而使飞行器上升。如果我们做一个简单的计算，就会发现，为了给典型的小型飞机提供所需的升力，机翼顶部的距离必须比底部的距离长约50%。图2-1显示了这样一个机翼的外观。

我们可以看到这个典型的小型飞机的机翼，它的顶面比底部长1.5%~2.5%，我们发现一个塞斯纳172将不得不以超过400英里（1英里≈1.609 3千米）每小时飞行，以产生足够的升力。显然，这种升降原理的描述是有缺陷的。

但是，谁说分离的空气必须同时在后缘相遇？图2-2显示了模拟风洞中机翼上的气流。在模拟中，定期引入彩色烟雾。人们可以看到，机翼顶部的空气比机翼下面的空气早到达后缘。事实上，仔细检查显示，机翼下方的空气在大气中的"自由流动"速度减慢。这便极大支持了等效转化时间原则。

普遍的解释也就是意味着颠倒飞行是不可能的。机翼在急速转弯时如何调整负载的巨大变化呢？那么为什么这个流行的解释持续了这么久？答案之一便是伯努利原理很容易理解。伯努利原理没有任何错误，也没有说空气在机翼顶部的速度更快。但是，正如上面的讨论所表明的那样，我们的理解并不完整。问题是当我们应用伯努利的原理时，我们忽略了一个重要的方面，如果我们知道机翼下方的空气速度，我们就可以计算机翼周围的压力。

2. 牛顿定律与升力

那么，机翼如何产生升降？要开始了解升降，我们必须回到高中物理学，并回顾牛顿的第一和第三定律。牛顿第一定律规定，任何物体在不受外力的情况下，将保持静止或匀速直线运动。这意味着，如果看到空气流出现弯曲，或者如果原来静止的空气被加速运动，则存在作用在其上的力。牛顿第三定律指出，对于每一个运动来说，都存在作用力与反作用力，且二者是等大反向的。举个例子，放在桌子上的物体对桌子施加一个力，那么桌子则对物体施加相等且方向相反的力以保持平衡。为了产生升力，机翼必须对空气做出响应。

我们来比较这两幅用来显示机翼上空气流（流线）的图片。在图2-3中，空气直接在机翼周围弯曲，然后直接落在机翼的后面。我们都在飞行手册中看过类似的图片。但是，当空气在飞机前面时，便完全离开了机翼。机翼没有对空气产生作用，因此没有产生升力！图2-4显示了空气流线，空气通过机翼，向下弯曲。机翼上的升力就是对空气弯曲的响应。

New Words and Expressions:

Newton ['nju:t(ə)n] *n*. 牛顿(英国科学家)；牛顿(力的单位)

Bernoulli [bə:'nu:li]*n*.伯努利（瑞士物理学家，数学家）

aeronautical [,eərə'nɔ:tɪkl]*adj*. 航空的；航空学的；飞机驾驶员的

simulation[,sɪmjʊ'leɪʃən]*n*.仿真；模拟；模仿；假装

capability [,keɪpə'bɪləti]*n*.才能，能力；性能，容量

geometric [,dʒɪə'metrɪk]*adj*. 几何学的；[数] 几何学图形的

converge[kən'vɜ:dʒ] *vi*.聚集；靠拢；收敛

free-stream 自由气流

streamline ['stri:mlaɪn] *n*. 流线；流线型

intuitive [ɪn'tju:ɪtɪv] *adj*.直觉的；凭直觉获知的

curve[kɜ:v] *n*.曲线；弯曲；曲线球；曲线图表

airfoil ['eəfɔɪl] *n.* 机翼；螺旋桨；翼型

airflow ['eəfləʊ] *n.* 气流（尤指飞机等产生的）；空气的流动

manual ['mænjʊ(ə)l] *n.* 手册，指南

Questions:

Answer the following questions according to the text.

（1）What is the Newton's laws?

（2）What is the popular explanation about "what makes an airplane fly"?

（3）What is the different between the fig.2-3 and fig.2-4 in the text?

Passage Two: The Principle of Flight (2)

1. The wing as a pump

As Newton's laws suggests, the wing must change something of the air to get lift. Changes in the air's momentum will result in forces on the wing. To generate lift a wing must divert lots of air down.

The lift of a wing is equal to the rate of change in momentum of the air it is diverting down. Momentum is the product of mass and velocity. The lift of a wing is proportional to the amount of air diverted down per second times the downward velocity of that air. It's that simple, for more lift the wing can either divert more air (mass) or increase its downward velocity. To the pilot the air is coming off the wing at roughly the angle of attack. To the observer on the ground, if he or she could see the air, it would be coming off the wing almost vertically. The greater the angle, the greater the vertical velocity. Likewise, for the same angle, the greater the speed of the wing the greater the vertical velocity. Both the increase in the speed and the increase of the angle increase the length of the vertical arrow. It is this vertical velocity that gives the wing lift. As stated, an observer on the ground would see the air going almost straight down behind the plane. This can be demonstrated by observing the column of air behind a propeller, a household fan, or under the rotors of a helicopter （see Fig.2-5）.

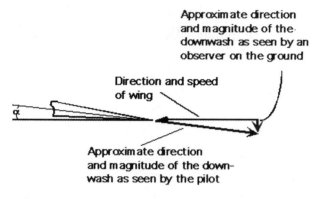

Fig.2-5   How downwash appears to a pilot and to an observer
on the ground

If we estimate that the average vertical component of the downwash of a Cessna 172 traveling at 110 knots, then to generate the needed 2,300 lbs of lift the wing pumps a whopping 2.5 ton/sec of air! The amount of air pumped down for a Boeing 747 to create lift for its roughly 800,000 pounds takeoff weight is incredible indeed.

Pumping, or diverting, so much air down is a strong argument against lift being just a surface effect as implied by the popular explanation. In fact, in order to pump 2.5 ton/sec the wing of the Cessna 172 must accelerate all of the air within 9 feet above the wing.

So how does a thin wing divert so much air? When the air is bent around the top of the wing, it pulls on the air above it accelerating that air down, otherwise there would be voids in the air left above the wing. Air is pulled from above to prevent voids. This pulling causes the pressure to become lower above the wing. It is the acceleration of the air above the wing in the downward direction that gives lift.

As the wing moves along, air is not only diverted down at the rear of the wing, but air is pulled up at the leading edge. This upwash actually contributes to lift and more air must be diverted down to compensate for it.

Normally, one looks at the air flowing over the wing in the frame of reference of the wing. In other words, to the pilot the air is moving and the wing is standing still. We have already stated that an observer on the ground would see the air coming off the wing almost vertically. But what is the air doing above and below the wing? Fig.2-6 shows an instantaneous snapshot of how air molecules are moving as a wing passes by. Remember in this figure the air is initially at rest and it is the wing moving. Ahead of the leading edge, air is moving up (upwash). At the trailing edge, air is diverted down (downwash). Over the top the air is accelerated towards the trailing edge. Underneath, the air is accelerated forward slightly.

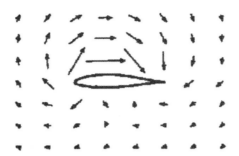

Fig.2-6　Direction of air movement around a
wing as seen by an observer on the ground

In the mathematical aerodynamics description of lift this rotation of the air around the wing gives rise to the "bound vortex" or "circulation" model. The advent of this model, and the complicated mathematical manipulations associated with it, leads to the direct understanding of forces on a wing. But, the mathematics required typically takes students in aerodynamics some time to master.

One observation that can be made from fig.2-6 is that the top surface of the wing does much more to move the air than the bottom. So the top is the more critical surface. Thus, airplanes can carry external stores, such as drop tanks, under the wings but not on top where they would interfere with lift. That is also why wing struts under the wing are common but struts on the top of the wing have been historically rare. A strut, or any obstruction, on the top of the wing would interfere with the lift.

2. The wing as air "scoop"

We now would like to introduce a new mental image of a wing. One is used to thinking of a wing as a thin blade that slices though the air and develops lift somewhat by magic. The new image that we would like you to adopt is that of the wing as a scoop diverting a certain amount of air from the horizontal to roughly the angle of attack, as depicted in fig.2-7. The scoop can be pictured as an invisible structure put on the wing at the factory. The length of the scoop is equal to the length of the wing and the height is somewhat related to the chord length (distance from the leading edge of the wing to the trailing edge). The amount of air intercepted by this scoop is proportional to the speed of the plane and the density of the air .

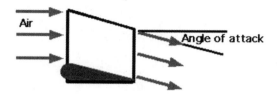

Fig.2-7　The wing as a scoop

As stated before, the lift of a wing is proportional to the amount of air diverted down times the vertical velocity of that air. As a plane increases speed, the scoop diverted more air. Since the load on the wing, which is the weight of the plane, does not increase the vertical speed of the diverted air must be decreased proportionately. Thus, the angle of attack is reduced to maintain a constant lift. When the plane goes higher, the air becomes less dense so the scoop diverts less air for the same speed. Thus, to compensate the angle of attack must be increased.

参考译文：飞行原理(2)

1. 机翼为泵

正如牛顿定律所描述的那样，机翼必须通过影响空气流动来获得升力。空气流动方向的变化将会导致机翼所受的力发生变化。为了产生升力，机翼必须将大量空气向下转移。

机翼的升力等于它周围流动空气的动量变化率。动量是质量和速度的乘积。机翼的升力与每秒向下流动的空气量成正比例。那么显然，若要获得更多的升力，机翼必须转移更多的空气（质量）或增加空气向下流动的速度。对于飞行员来说，空气从机翼上脱离的角度大致是迎角的角度。对地面上的观察员来说，他（她）所看到的空气将几乎垂直离开机翼。角度越大，垂直速度就越大。同样，若角度相同，那么机翼的速度越大，垂直速度就越大。速度的增加和迎角的增加能增加垂直箭头的长度。正是这种垂直速度给机翼提供了升力。如上所述，观察员在地面上会看到空气在飞机后直线直落。这可以通过观察螺旋桨、家用风扇或直升机旋翼所产生的空气柱来证明（见图2-5）。

如果我们估计以110海里每小时的速度飞行的塞斯纳172下冲的平均垂直分量，那么为了产生所需的2300磅（1磅≈0.4536千克）升力，机翼吸引空气的速度必须达到2.5吨/秒！使波音747起飞的空气量大约为80万磅，这是不可思议的。

产生升力需要吸转这么多的空气只是普遍解释给出的一个表层效果。事实上，为了使吸出2.5吨/秒的空气，塞斯纳172号必须使机翼上方9英尺（1英尺≈0.304 8米）内的所有空气都加速流动。

那么薄翼如何转移这么多的空气呢？当空气绕着机翼的顶部弯曲流动时，它将使其上方的空气加速向下流动，否则在机翼上方的空气中将会有空隙。上方的空气流动以防止出现空隙。这种流动使机翼上方的压力变得低于下方。这使机翼上方的空气在向下加速流动产生上升力。

当机翼移动时，空气不仅在机翼的后部向下流动，而且机翼前缘的空气会向上流动。这个上升流实际上有助于提升，更多的空气必须被向下转移以补偿流失的空气。

通常，在机翼的参考框架中，可以看到在机翼上流动的空气。换句话说，从飞行员的角度看，空气正在移动，机翼静止不动。我们已经说过，地面上的一名观察员会看到空气几乎垂直脱离机翼落下，但机翼上方和下方的空气分别是什么样的？图2-6显示了空气分子在机翼经过时如何移动的瞬时快照。记住在这张图中，空气最初处于静止状态，是机翼在移动。在机翼前缘，空气正在向上流动。在后缘，空气被转向下方。在顶部，空气朝向后缘加速。在下面，空气会稍微向前加速。

在数学空气动力学描述中，机翼周围空气的旋转产生了"束缚涡流"或"循环"模型。这种模型的出现以及与之相关的复杂的数学操作使我们能够直接了解机翼上的力。但是，所需的数学知识通常需要我们掌握空气动力学才可以。

可以从图2-6中得到的一个观察结果是，机翼的顶面比底部空气流动量要大得多，因此顶部更为关键。因此，飞机可以在机翼下面装载外部设备，例如下水箱，但不能在顶部装载，否则会干扰飞机升降。这也是为什么在历史上翼下的翼支架是常见的，但翼顶部的翼支架是罕见的。机翼顶部的支柱或任何障碍物都将干扰飞机升降。

2. 机翼为"舀"

我们现在将机翼做一个新的比喻。将一个翼比作一个薄薄的刀片，对空气进行切割，并使其继而产生魔幻的升降效果。另一个新的比喻就是将机翼作为一个勺子，将一定数量的空气从水平转向大致的迎角，如图2-7所示。勺子可以被看作是一个存在于机翼上却看不见的结构。勺子的长度等于翼的长度，并且高度与弦长度（从翼的前缘到后缘的距离）相关。这个勺子拦截的空气量与飞机的速度和空气的密度成正比例。

如前所述，机翼的升降力与转移的空气量乘以空气的垂直速度成正比。当一架飞机速度增加时，"勺子"会转移更多的空气。由于机翼上的负载（即飞机质量）不会增加，那么空气的垂直速度必须按比例降低。因此，迎角降低以保持恒定的升力。当飞机越飞越高时，空气变得不那么密集，因此"勺子"以相同的速度转移较少的空气。因此，补偿迎角必须增加。

New Words and Expressions:

momentum [mə'mentəm]*n.*势头；[物] 动量；动力；冲力

velocity[və'la:səti]*n.*[物] 速度

diverted [daɪ'vɜːt; dɪ-]*vt.* 转移；使……欢娱；使……转向

acceleration [əkselə'reɪʃ(ə)n]*n.* 加速，促进；[物] 加速度

compensate ['kɒmpenseɪt]*vi.*补偿，赔偿；抵消

upwash['ʌpwɒʃ]*n.*上洗流；升流；上升风；[航] 气流上洗

downwash['daʊnwɒʃ]*n.*气流下洗；向下运动；由高处冲下的物质

vertically['və:tɪkli]*adv.*垂直地

column ['kɒləm]*n.* 纵队，列；专栏；圆柱，柱形物

propeller [prə'pelə]*n.* [航][船] 螺旋桨；推进器

rotor ['rəʊtə]*n.* [电][机][动力] 转子；水平旋翼；旋转体

helicopter ['helɪkɒptə]*n.*[航] 直升飞机

frame [freɪm]*n.*框架；结构；[电影] 画面

instantaneous [ˌɪnst(ə)n'teɪnɪəs]*adj.* 瞬间的；即时的；猝发的

snapshot ['snæpʃɒt]*n.* 快照，快相；急射，速射；简单印象

mathematical[mæθ(ə)'mætɪk(ə)l]*adj.* 数学的，数学上的；精确的

aerodynamics[ˌeərə(ʊ)daɪ'næmɪks]*n.* [流] 气体力学；[航] 航空动力学

rotation [rə(ʊ)'teɪʃ(ə)n]*n.*旋转；循环，轮流

bound vortex 约束涡；附着涡；束缚涡

proportional [prə'pɔːʃ(ə)n(ə)l]*adj.* 比例的，成比例的；相称的，均衡的

Questions:

Answer the following questions according to the text.

（1）How does a thin wing divert so much air?

（2）What gives the wing lift?

（3）What is the difference between the air flow in different directions of the wings?

Passage Three: The Principle of Flight (3)

1. Lift requires power

When a plane passes overhead the formerly still air ends up with a downward velocity. Thus, the air is left in motion after the plane leaves. The air has been given energy. Power is energy, or work, per time. So, lift must require power. This power is supplied by the airplane's engine (or by gravity for a sailplane).

How much power will we need to fly? The power needed for lift is the work (energy) per unit time and so is proportional to the amount of air diverted down times the velocity squared of that diverted air. We have already stated that the lift of a wing is proportional to the amount of air diverted down times the downward velocity of that air. Thus, the power needed to lift the airplane is proportional to the load (or weight) times the vertical velocity of the air. If the speed of the plane is doubled, the angle of attack must be reduced to give a vertical velocity that is half the original to give same lift. This shows that the power required for lift becomes less as the airplane's speed increases. In fact, we have shown that this power to create lift is proportional to one over the speed of the plane.

We all know that to go faster ,we must apply more power. So there must be more to power than the power required for lift. The power associated with lift, described above, is often called the "induced" power. Power is also needed to overcome what is called "parasitic" drag, which is the drag associated with moving the wheels, struts, antenna, etc. through the air. The energy the airplane imparts to an air molecule on impact is proportional to the speed squared. The number of molecules struck per time is proportional to the speed. Thus the parasitic power required to overcome parasitic drag increases as the speed cubed. At cruise, the power requirement is dominated by parasitic power.

There is a misconception by some that lift does not require power. This comes from aeronautics in the study of the idealized theory of wing sections (airfoils). Since we have seen that the power necessary for lift is proportional to one over the length of the wing, a wing of infinite span does not require power for lift. If lift did not require power airplanes would have the same range full as they do empty, and helicopters could hover at any altitude and load.

2. Power and wing loading

Let us now consider the relationship between wing loading and power. Does it take more power to fly more passengers and cargo? And, does loading affect stall speed? At a constant speed, if the wing loading is increased the vertical velocity must be increased to compensate. This is done by increasing the angle of attack. If the total weight of the airplane were doubled, the vertical velocity of the air is doubled to compensate for the increased wing loading. The induced power is proportional to the load times the vertical velocity of the diverted air, which have both doubled. Thus the induced power requirement has increased by a factor of four! The same thing would be true if the airplane's weight were doubled by adding more fuel, etc.

One way to measure the total power is to look at the rate of fuel consumption. Fig.2-8 shows the fuel consumption versus gross weight for a large transport airplane traveling at a constant speed. Since the speed is constant the change in fuel consumption is due to the change in induced power. The data are fitted by a constant (parasitic power) and a term that goes as the load squared. This second term is just what was predicted in our Newtonian discussion of the effect of load on induced power.

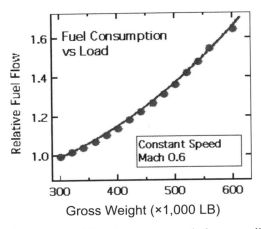

Fig.2-8　Fuel consumption versus load for a large transport airplane traveling at a constant speed

3. Wing vortices

One might ask what the downwash from a wing looks like. The downwash is related to the details on the load distribution on the wing. The distribution of load changes from the root of the wing to the tip, the amount of air in the downwash must also change along the wing. The wing near the root is "scooping" up much more air than the tip. Since the root is diverting so much air，the net effect is that the downwash will begin to curl outward around itself, just as the air bends around the top of the wing because of the change in the velocity of the air. This is the wing vortex. The tightness of the curling of the wing vortex is proportional to the rate of change in lift along the wing. At the wing tip the lift rapidly become zero causing the tightest curl. This is the wing tip vortex and is just a small part of the wing vortex.

Winglets (those small vertical extensions on the tips of some wings) are used to improve the efficiency of the wing by increasing the effective length of the wing. The lift of a normal wing must go to zero at the tip because the bottom and the top communicate around the end. The winglets blocks this communication so the lift can extend farther out on the wing. Since the efficiency of a wing increases with length, this gives increased efficiency. One caveat is that winglet design is tricky and winglets can actually be detrimental if not properly designed.

参考译文：飞行原理(3)

1. 升降需要做功

当一架飞机从头顶飞过时，此处本来静止的空气开始向下不断流动。因此，当飞机离开之后空气仍然保持运动，空气被赋予了能量。功率是单位时间的能量或者做功。因此，升降必须要有动力支持。这种动力由飞机的发动机提供（对滑翔机而言为重力能）。

飞机飞行所需要的功率为多少呢？升降所需的功率是单位时间做的功（能），因此它与空气流动量和速度的平方之积成正比。我们已经指出，机翼的升力与空气转移量和空气的下降速度之积成正比。因此，飞机飞行所需的功率与载荷（重力）和垂直流动速度空气之积成正比。如果飞机的速度加倍，那么，迎角必须减小，以提供原始垂直速度的一半从而提供相同的升力。这表明升力所需的功率随着飞机速度的增加而变小。但事实上，升降功率是与飞机的速度成正比的。

我们都知道要想使飞机飞得更快，我们必须给予它更多的功率。因此提供给飞机飞行的动力必须比所需的升降动力更多。如上所述，与升降相关的动力通常被称为"诱导"动力。另外还需要动力来克服所谓的"寄生"阻力，这是与空气移动飞行器、支柱、天线等穿越大气相关的阻力。飞机在撞击空气分子时施加的能量与速度平方成正比，每次撞击的分子数与速度成比例。因此，克服寄生阻力所需的寄生力随着速度的立方的增加而增加。在飞行器巡航时，功率需求由寄生功率决定。

有些人错误地认为升降不需要功率，这来自航空学研究翼型所给的理想化理论。由于我们已经知道飞机升降所需的功率与翼的长度成正比，无限跨度的翼片在升降时不需要力量。如果升降不需要功率，飞机将占据整个空间范围，直升机可以在任何高度和负载下悬停。

2. 动力和机翼装载

现在让我们来考虑机翼装载与动力之间的关系。飞机若搭载更多乘客和货物就一定需要更多的功率吗？另外，超载会影响失速速度吗？若速度恒定，机翼负载增加，那么垂直速度必须增加以补偿。这是通过增加攻角来实现的。如果飞机的总重量加倍，那么空气的垂直速度就要加倍，以补偿增加的机翼载荷。感应功率与负载和转向空气的垂直速度之积成正比，两者都加倍。因此，感应功率需求增加了四倍！如果燃料增加使飞机的质量加倍，那么也是如此。

测量总功率的方法之一是查看燃油消耗率。图2-8显示以恒定速度行驶的大型运输飞机的燃料消耗与总重量的比例关系。由于速度恒定，燃油消耗的变化是由于感应功率的变化引起的。数据由常数（寄生功率）和负载平方项拟合。第二个术语就是利用牛顿定律所讨论的负载对感应功率的影响。

3. 翼旋涡

人们可能会问机翼的向下运动是什么样的。向下运动与机翼上的负载分布有关。负载分布从机翼的根部到尖端是有变化，向下运动的空气量也必须沿着翼变化。机翼根部会比尖端"舀起"更多的空气。由于根部转移了很多空气，效果会是向下运动气流将自动向外卷曲，就像空气由于空气速度的变化而绕着机翼的顶部弯曲一样，这就是翼涡。机翼涡流卷曲的紧密程度与沿机翼升力的变化率成比例。在翼尖，升力迅速变为零，导致机翼卷曲到最紧。这是翼尖涡流，只是翼涡的一小部分。

小翼（翼尖端上的小垂直延伸部分）用于通过增加机翼的有效长度来提高机翼的效率。因为底部和顶部在末端连通，所以正常翼的升力必须在尖端处为零。小翼阻挡了这个连通，所以机翼可以获得更大的伸展范围。由于翼的效率随着长度的增加而增加，所以机翼越长，效率提高。一个值得注意的问题是，小翼设计是棘手的，如果设计不正确，小翼的存在实际上可能是有害的。

New Words and Expressions:

downward ['daʊnwəd]*adj.*向下的，下降的

gravity ['ɡrævətɪ]*n.*重力，地心引力；严重性；庄严

thermal['θɜːm(ə)l]*n.*上升的热气流

sailplane['seɪlpleɪn]*n.*滑翔机

proportional [prə'pɔːʃ(ə)n(ə)l]*adj.*比例的，成比例的；相称的，均衡的

induced[ɪn'djuːst]*adj.*感应的；诱发型

parasitic [pærə'sɪtɪk] *adj.*寄生的

antenna [æn'tenə] *n.*[电信] 天线；[动] 触角，[昆] 触须

molecule ['mɒlɪkjuːl] *n.*[化学] 分子；微小颗粒，微粒

cruise [kruːz] *vi.*巡航，巡游；漫游

misconception [mɪskən'sepʃ(ə)n] *n.*误解；错觉；错误想法

aeronautics[eərə'nɔːtɪks] *n.*航空学；飞行术

wing section 翼截面；[航] 翼剖面；翼型

cargo ['kɑːgəʊ] *n.*货物，船货

fuel consumption 耗油量

parasitic power[电] 寄生功率

Newtonian[njuː'təuniən]*adj.*牛顿学说的

wing vortices 翼旋涡

winglets ['wɪŋlɪt] *n.*小翅，小翼

vertical extensions 垂直蔓延，纵向延伸

Questions:

Answer the following questions according to the text.

（1）How much power will we need to fly?

（2）Does the wing take more power to fly more passengers and cargo?

（3）How does loading affect stall speed?

Unit Three

Passage One: CAD/CAM and Aircraft Design

Time compression strategies in a globally competitive environment in the development of product have become one of the crucial factors to be considered. This led to the integration of computer technology in design and manufacturing of products. CAD/CAM (computer-aided design and computer-aided manufacturing) refers to computer software that aids in both designing and manufacturing of the products. Also the CAD software enables creating of various computer models that are defined by geometrical parameters whereas CAM uses the geometrically designed data to control the automated machinery systems. These software programs are used in various fields but there seems no industry in which it is extremely essential than the aerospace industry as it requires the innovations blended with highest order of complexity.

With the use of computer aided design, the performance of various structures that are subjected to static and fluctuating loads can now be easily simulated, rapidly analyzed and tested in more efficient and precise manner. The two-engine Boeing 777 passenger airplane, for example, was completely designed by computer design i.e. paperless design, with 2,000 workstations linked to eight design servers (computers). Design can be optimized and also rapid modifications can be made directly at any time ensuring a reliable and quality design. CAD system accurately and quickly produces the models definitions for the products and their components of higher quality and better consistency than those produced by traditional manual drafting.

CAM refers to the use of a computer to assist in all manufacturing plant operations, including planning, management, transportation and storage. Also it allows creating of components in rapid manner and tooling with more precise and accurate dimensions. CAM software employs the models and assemblies that are designed in CAD software describing the tool path for various operations which drive the numeric control machines that turn the designs into physical components/parts. These are most often used in machining operations of finished components. Thus, the emergence of CAD/CAM does have a major impact on manufacturing, by standardizing product development.

CAD/CAM software is compatible with most common file formats to give you the flexibility to accept a wide variety of file types from your clients. The CAD Design Tree acts as a history of your model creation and allows you to go back to any step in your process to make modifications to your design.CAD allows you to define and organize groups of geometry using layers that can be easily turned on and off to give you enhanced design and viewing control of your model. The Select By Layer functionality offers a convenient means of selecting sections of your geometry quickly and efficiently.

Whether you designed or imported a client's file, CAD/CAM makes it easy to measure geometry, create part prints, and generate inspection reports so you can easily understand and communicate part geometry.

Cross Section Viewing allows you to inspect the internal details of your solid model as if it were cut straight across any plane. This powerful viewing feature also gives you the ability to generate wireframe geometry of the cross section.

Dynamic drawing offers the control and flexibility needed to easily create your part geometry. This powerful feature gives you the ability to push and pull surfaces and solid model geometry to quickly adjust part dimensions. It also allows you to easily move and rotate your parts around the workspace, which helps to speed up the design process.

All dynamic drawing functions include a snap increment function, which acts as a uniform grid in the workspace. This allows you to define snap increments based on distances or angles for rotation functions. This feature creates a whole new level of efficiency for dynamic drawing by greatly reducing the number of clicks and data entry modifications.

CAD/CAM offers a wide variety of easy to use 2D geometry tools such as Points, Lines, Arcs, Splines, Offsets and Parallel Lines to sketch your parts. You'll also have access to commonly used wireframe shapes such as Rectangles, Ellipses, Gears, Cams, Sprockets and Bolt Hole Patterns.

The CAD design software provides 40 preprogrammed commonly used fabrication shapes that can be easily modified to fit the specific needs of the design.Easily add text to your drawings with tools like Text Along A Curve, Extrude Text and Project To Curve for 3D surfaces. The software allows you to work with standard Windows fonts & open fonts or select from a library of CAD/CAM fonts.The CAD software offers powerful wireframe editing tools to give you the control needed to easily manipulate part geometry using functions for scaling, trimming, moving, rotating and more.CAD delivers robust solid model editing tools like Extrude, Split, Shell, Booleans and more, so you easily can manipulate your models for fast and efficient 3D part creation.

参考译文：CAD／CAM与飞机设计

全球竞争环境中产品开发中的时间压缩策略已经成为要考虑的关键因素之一。这导致了计算机技术在产品设计和制造中的结合。CAD／CAM（计算机辅助设计和计算机辅助制造）是指帮助设计和制造产品的计算机软件。CAD软件还可以创建由几何参数定义的各种计算机模型，而CAM使用几何设计的数据来控制自动化机械系统。这些软件程序用于各种领域，但由于航空航天工业需要高复杂度的创新，因此CAD/CAM在该行业的重要性比其他行业都高。

通过使用计算机辅助设计，如今可以更有效和精确地模拟，并快速分析和测试受到静态和波动负载的各种结构的性能。例如，双引擎波音777客机完全由计算机设计，即无纸设计，2 000台工作站与八台设计服务器（计算机）相连。可以优化设计，并可以随时直接进行快速修改，以确保可靠和高质量的设计。与传统的手工绘图相比，CAD系统可以准确、快速地为产品及其组件提供更高质量和更高协调性的模型定义。

CAM是指使用计算机来协助制造工厂的所有运营，包括规划、管理、运输和存储。此外，它可以以快速创建模具组件，并具有更精确和准确的尺寸。CAM软件采用在CAD软件中设计的模型和装配件，描述各种操作的刀具路径，驱动数控系统将设计模型转变为实体部件。这些都是工件最常用的加工操作。因此，CAD／CAM的出现确实对飞行器制造业的标准化生产发展有重大影响。

CAD/CAM软件能与大多数常见的文件格式兼容，使您可以灵活地接受来自客户端的各种文件类型。CAD设计树会对创建模型的过程进行记录，并且可以返回到设计过程中的任何步骤，以对设计方案进行修改。CAD允许使用可以轻松打开和关闭的图层来定义和组织几何组，从而为增强模型的设计和查看控制提供了便利。"按层选择"功能提供了一种方便的方式来快速有效地选择几何图形。

无论您设计或导入客户端的文件，CAD/CAM都可以轻松测量几何图形，创建零件打印件，并生成检查报告，以便您轻松了解和传达几何零件。

交叉剖视图允许直线切割任何平面以便检查实体模型的内部细节。这种强大的查看功能还可以生成横截面的几何线框。

动态绘图提供了轻松创建零件几何所需的可操控性和灵活性。这个强大的功能能够拉伸表面和实体模型几何，以快速调整零件尺寸。它还允许您轻松地在工作区周围移动和旋转零件，这有助于加快设计过程。

所有动态绘图功能都包括一个增量捕捉功能，它在工作空间中表现为一个均匀的网格。这允许您根据旋转的距离或角度定义捕捉增量。此功能大大减少点击次数和数据输入修改次数，将动态绘制提高到了一个新的效率水平。

CAD/CAM提供了各种易于使用的2D几何工具来绘制零件，例如点、线、弧、样条、偏移和平行线。您还可以使用各种常用的线框形状，如矩形、椭圆、齿轮、凸轮、链轮和螺栓孔等。

CAD设计软件提供40多种预编程用的制造形状，可以轻松修改以满足设计的特定需求。使用"曲线文本""拉伸文本"和"3D曲面曲线"等工具轻松添加文本。软件允许您使用标准的Windows字体及其他可使用的字体，或从CAD/CAM字体库中进行选择。CAD软件提供强大的线框编辑工具，为您提供缩放、修剪、移动、旋转等功能，以便您轻松完成操作零件几何所需的控制。CAD提供了强大的实体模型编辑工具，如拉伸、分解、壳、布尔运算等，可以使您轻松操纵模型以便快速有效地完成3D部件创建。

New Words and Expressions:

context['kɒntekst]*n.*环境；上下文；来龙去脉

strategy['strætədʒɪ]*n.*战略，策略

geometrical[dʒɪə'metrɪkl]*adj.*几何的，几何学的

parameter [pə'ræmɪtə]*n.*参数；系数；参量

machinery[mə'ʃi:n(ə)rɪ]*n.* 机械；机器；机构；机械装置

conceptualize[kən'septjʊəlaɪz]*vt.*使概念化

prototype ['prəʊtətaɪp]*n.*原型；标准，模范

storage ['stɔːrɪdʒ]*n.*存储；仓库；贮藏

dimensioning[dɪ'mɛnʃən]*n.*尺寸标注；量尺寸；标示尺寸

inspection [ɪn'spekʃn] *n.*视察，检查

wireframe ['waɪəfreim] *n.*线框

uniform grid   均匀网格

parallel lines   平行线

sketch[sketʃ] *n.*素描；略图；梗概

rectangle[rɛk'tæŋgl] *n.* [数] 矩形；长方形

ellipse [ɪ'lɪps] *n.*[数] 椭圆形，[数] 椭圆

gear [gɪə] *n.* 齿轮；装置，工具；传动装置

sprocket ['sprɒkɪt] *n.*链轮齿；扣链齿轮

boolean ['buːlian] *adj.* 布尔数学体系的

cylinder ['sɪlɪndə] *n.*圆筒；汽缸；[数] 柱面；圆柱状物

Questions:

Answer the following questions according to the text.

（1）What is CAD/CAM ?

（2）What is the function of the wireframe editing tool?

（3）What is the role of the sectional view?

Passage Two: FEM and Aircraft Design

Computer-aided engineering analysis based on the finite element (FE) method is recognized to be a very effective numerical simulation and optimization technique in the field of aircraft design. It has important guiding significance and practical value that can improve product quality and performance, reduce production costs, shorten design cycle, and so on.

FE modeling is data pre-processing for FE analysis. The calculation accuracy of the FE method depends on the degree of approximation of physical characteristics of the model and its real structure. Therefore, establishing a correct and reasonable FE model is the most important issue to carry out FE analysis and optimization. However, there is an encountered problem that a significant amount of time and workforce is required for creation and modification of the FE model. In addition, it is one of the major difficulties in application of the FE method. Therefore, high-quality and automated FE modeling has been an important direction of research on aircraft structural FE technique.

The current process of FE modeling of aircraft wing structures is shown in Fig.3-1. Firstly, geometric model is generated by the CATIA system manually, and then imported in PATRAN. Secondly, FE meshing and property loading are usually completed via manual operations. Therefore, as most operations of FE modeling are interactive, design quality and results rely heavily on technical level and experience of designers while lack effective means of digital design knowledge and experience accumulated in

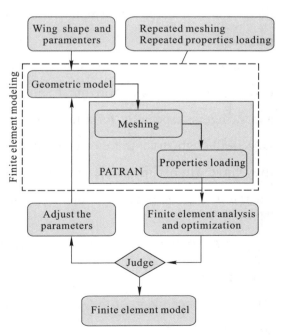

Fig.3-1   Process of finite element modeling of aircraft wing structure

the design process. In addition, because of the complexity of wing structure, the interactive approach is difficult to meet the demand for rapid modeling during design.

The layout of wing structure, geometric mesh model and FE model including property data are established on parametric description and automatic update. The generation processes of skeleton model, geometric mesh model, and FE model of wing structure are achieved based on CAD parametric technology. The method settles a series of problems of geometric model description, parameter association and model automatic update in the process of FE modeling which establishes a key technical basis for parametric FE analysis and optimization.

After rapid FE modeling completion, the definition of the aircraft structural preliminary design is formed, and the final FE analysis model and the optimized size parameters of the main components can be output.

1. Knowledge-based template parametric technique

Template can be considered as a technique similar to knowledge reuse in view of case reasoning, which is based on thing similarity and multiplexing principle of design methods. The basic idea is that the thing extracted from a class of analogous things and any similar things can be deemed as a template. The core of template is reuse of design information and parametric variations.

Internal components of aircraft wing are not greatly different in spite of numerous parts, such as wing and horizontal tail. Some of them are composed by skins, spars, ribs, and stringers, especially for a high aspect ratio wing used on no matter an airliner or an UAV. Although structural dimension and the number of components are usually different, they have almost the same function, layout and generation process of components. Therefore, the components of aircraft wings have a strong template feature.

2. Parametric layout of wing structure

Structural design of an aircraft wing requires a layout of spars, ribs and stringers based on the reference plane of the wing and generation of a skeleton model of wing structure. Skeleton model, the backbone of the model design, is an assembly design model with multi-rank abstract levels, signified with geometric elements of points, lines, surfaces and all kinds of benchmarks in CATIA. Skeleton model includes planes, axes and other structure layout information of components.

3. Parametric design of geometric mesh model

Geometric mesh model as a discrete form of skeleton model is in an important status in FE modeling. A connection is established between skeleton model and FE model. Based on geometric mesh model, properties of each element can be individually defined and viewed as optimization variables and check objects in the FE analysis and optimization.

Compatibility is required for FE mesh of an aircraft wing structure. If two components cross, on the line of the intersection, in order to ensure that the right combination of adjacent curved surface patches mesh in combinatorial surfaces, it is necessary to guarantee the same discrete form on common boundary and FE mesh to be completely consistent. On the line of the intersection, that is to say, nodes of the FE mesh of two components must be coincident. Therefore, in order to set up an FE model, starting from the untrimmed geometry, skeleton model produced by the structural design department needs mesh discretization.

4. Generation and description method of FE model

FE model comprises FE mesh and properties including physical properties and material properties. It contains all geometric and topological information of geometric mesh model. Meshing is conducted in geometric mesh model elements to ensure that the adjacent element mesh is consistent exactly on the common boundary. Properties of the components are stored in geometric mesh model, and element properties in FE model inherit directly from geometric mesh model. Every time that a reconstruction of FE model is needed, the mesh properties update, so that the corresponding element properties in FE model are always correct when the geometry changes.

A finite element parametric modeling method of aircraft wing structures is proposed because of time-consuming characteristics of finite element analysis pre-processing. The main research is positioned during the preliminary design phase of aircraft structures. A knowledge-driven system of fast finite element modeling is built. Based on this method, employing a template parametric technique, knowledge including design methods, rules and expert experience in the process of modeling is encapsulated and a finite element model is established automatically, which greatly improves the speed, accuracy and standardization degree of modeling. Skeleton model, geometric mesh model, and finite element model including finite element mesh and property data are established on parametric description and automatic update.

参考译文：有限元与飞机设计

基于有限元（FE）方法的计算机辅助工程分析被认为是飞机设计领域非常有效的数值模拟和优化技术。其具有提高产品质量和性能，降低生产成本，缩短设计周期等重要指导意义和实用价值。

有限元建模是指有限元分析的数据前处理。有限元方法的计算精度取决于模型的物理特性及其实际结构的近似程度。因此，建立正确合理的有限元模型是进行有限元分析和优化的最重要问题。然而，面临的问题是需要大量的时间和人力来创建和修改有限元模型。此外，它也是应用有限元技术的主要困难之一。因此，高质量和自动化的有限元模型一直是飞机结构有限元技术研究的重要方向。

目前，飞机机翼结构的有限元建模的过程如图3-1所示。首先，几何模型由CATIA系统手动生成，然后在PATRAN中导入。其次，FE网格划分和属性加载通常通过手动操作完成。因此，由于有限元建模的大多数操作是交互式的，设计质量和结果严重依赖于设计师的技术水平和经验，导致设计过程中缺乏积累数字设计知识和经验的有效方法。另外，由于机翼结构的复杂性，交互式方法难以满足设计过程中快速建模的需求。

机翼结构布局、几何网格模型和包含属性数据的有限元模型可以通过在参数描述和自动更新中建立。基于CAD参数技术可以实现骨架模型、几何网格模型和机翼结构有限元模型的生成过程。该方法在有限元建模过程中解决了几何模型描述、参数关联和模型自动更新等一系列问题，为有限元分析和优化提供了重要的技术依据。

快速有限元建模完成后，形成了飞机结构初步设计的定义，可以输出最终的有限元分析模型和主要部件的优化尺寸参数。

1. 基于知识的模块参数化技术

案例推理是指事物的相似性和设计方法的复用原理，因此模板可以被认为是类似于知识重用的技术。基本思想是指从一类类似的东西和任何类似的东西中提取的事物可以被认为是模板。模板的核心是设计信息和参数变化的再利用。

飞机机翼的内部部件的许多部件并没有太大差异，如机翼和水平尾翼。其中一些由蒙皮、翼梁、肋条和桁条组成，尤其是对于无论是客机或者无人机中使用的大展弦比机翼。虽然结构尺寸和组件数量通常不同，但它们具有几乎相同的功能、布局和组件的生成过程。因此，飞机机翼的部件具有很强的模板特征。

2. 机翼结构参数布局

飞机机翼的结构设计是指基于机翼的参考平面和机翼结构骨架模型生成的翼梁、肋条和桁条的布局。骨架模型是模型设计的主干，是具有多级抽象层次的装配设计模型，是以CATIA中的点、线、面和几何基准为基础。骨架模型包括平面、轴和组件的其他结构布局信息。

3. 几何网格模型的参数化设计

几何网格模型作为离散形式的骨架模型在有限元建模中处于重要地位。在骨架模型和有限元模型之间建立了连接。基于几何网格模型，每个单元的属性可以单独定义并可被视为优化变量，是有限元分析和优化中检查的对象。

飞机机翼结构的有限元网格需要兼容性。如果两个部件交叉，在交叉线上，为了确保相邻曲面在结合表面上正确组合网格，必须保证共同边界上离散形式相同且有限元网格完全一致。也就是说，在交叉线上两个组件的有限元网格的节点必须一致。因此，为了建立有限元模型，从未修剪几何开始，结构设计部门生成的骨架模型就需要网格离散化。

4. 有限元模型的生成和描述方法

有限元模型包括有限元网格、物理属性及材料属性。它包含了几何网格模型的所有几何信息和拓扑信息。网格划分在几何网格模型元素中进行，以确保相邻元素网格在公共边界上完全一致。部件的属性存储在几何网格模型中，有限元模型中的元素属性直接从几何网格模型中继承。每次需要重建有限元模型时，网格属性更新，这样使得几何模型变化时有限元模型中的相应单元格属性总是正确的。

鉴于有限元分析前处理费时的缺点，提出了飞机机翼结构的有限元参数建模方法。研究工作主要用于飞机结构初步设计阶段。基于知识驱动系统建立快速有限元建模方法，该技术采用模板参数化，通过封装建模过程中的设计方法、规则和专家经验等知识自动建立了有限元模型，大大提高了建模的速度、准确性和标准化程度。格架模型、几何网格模型和包括有限元网格和属性数据的有限元模型，通过参数描述和自动更新数值而建立。

New Words and Expressions:

parametric[ˌpærə'metrɪk]*adj*. [数][物] 参数的；[数][物] 参量的

time-consuming ['taimkənˌsjuːmiŋ]*adj*.耗时的；旷日持久的

accuracy ['ækjʊrəsɪ]*n*. [数] 精确度，准确性

parametric [ˌpærə'metrɪk]*adj*. [数][物] 参数的；[数][物] 参量的

optimization [ˌɒptɪmaɪ'zeɪʃən] [ˌɒptɪmaɪ'zeɪʃən] *n* .优化；最佳化

calculation [ˌkælkjʊ'leɪʃ(ə)n]*n*. 计算；估计；计算的结果；深思熟虑

approximation [əˌprɒksɪ'meɪʃn]*n*. [数] 近似法；接近；[数] 近似值

Iteration [ɪtə'reɪʃ(ə)n]*n*. [数] 迭代；反复；重复

association [əˌsəʊsɪ'eɪʃn]*n*. 协会，联盟，社团；联合；联想

preliminary [prɪ'lɪmɪn(ə)rɪ]*n*. 准备；预赛；初步措施

analogous [ə'næləgəs]*adj*. 类似的；[昆] 同功的；可比拟的

benchmarks[bɛntʃˌmɑːk]*n*. [计] 基准；水准点；基准测试程序数值（benchmark的复数形式）

combinatorial [kɒmˌbaɪnə'tɔːrɪəl]*adj*. 组合的

Questions:

Answer the following questions according to the text.

（1）What is the finite element modeling method proposed in this article?

（2）How is finite element analysis applied in wing structure design?

（3）How to design a geometric network model?

Unit Four

Passage One: Airplane Construction and Control

Aircraft and spacecraft fly in totally different environments, so they need different methods to direct their movement and to maintain their orientation.

To provide stability and control, most airplanes use various control surfaces that work on the same principle as a wing, while spacecraft use thrust and spin.

How does a pilot control an airplane?Moveable surfaces on an airplane's wings and tail allow a pilot to maneuver an airplane and control its attitude or orientation. These control surfaces work on the same principle as lift on a wing. They create a difference in air pressure to produce a force on the airplane in a desired direction.

The main purpose of the tail is to provide stability. If tilted by a gust of wind, a stable airplane tends to recover, just as a ball lying at the bottom of a bowl will roll back to the center after being disturbed.

What is an airplane's "tail fin" for?A vertical stabilizer, or tail fin, keeps the airplane lined up with its direction of motion. Air presses against both its surfaces with equal force when the airplane is moving straight ahead. But if the airplane pivots to the right or left, air pressure increases on one side of the stabilizer and decreases on the other. This imbalance in pressure pushes the tail back into line.

What is an airplane's "rear wing" for?Like the vertical stabilizer, the horizontal stabilizer helps keep the airplane aligned with its direction of motion. If the airplane tilts up or down, air pressure increases on one side of the stabilizer and decreases on the other, pushing it back to its original position. The stabilizer also holds the tail down, a result of the airplane's center of gravity being forward of the wing's center of lift.

How is controlling an airplane different than controlling car or boat? Stability and control are much more complex for an airplane, which can move freely in three dimensions, than for cars or boats, which only move in two. A change in any one of the three types of motion affects the other two.

Imagine three lines running through an airplane and intersecting at right angles at the airplane's center of gravity.

(1) Rotation around the front-to-back axis is called roll.

(2) Rotation around the side-to-side axis is called pitch.

(3) Rotation around the vertical axis is called yaw.

1. The Ailerons Control Roll

On the outer rear edge of each wing, the two ailerons move in opposite directions, up and down, decreasing lift on one wing while increasing it on the other. This causes the airplane to roll to the left or right. To turn the airplane, the pilot uses the ailerons to tilt the wings in the desired direction.

2. The Elevator Controls Pitch

On the horizontal tail surface, the elevator tilts up or down, decreasing or increasing lift on the tail. This tilts the airplane up and down.

3. The Rudder Controls Yaw

On the vertical tail fin, the rudder swivels from side to side, pushing the tail in a left or right direction. A pilot usually uses the rudder along with the ailerons to turn the airplane.

How do you control attitude in space?Because the control surfaces that an airplane uses cannot work in airless space, a spacecraft relies on a different form of attitude control. To change orientation, a spacecraft applies torque (a twisting force) by firing small rockets or by spinning internal wheels.

Some satellites use a system of reaction wheels—basically gyroscopes—to change their attitude in space. The rapidly spinning wheels carry a lot of rotational momentum that the spacecraft can tap to change its own orientation. To do this in all three dimensions of space, the spacecraft must use three rotating wheels oriented at right angles to one another.

How can thrust help change attitude?Most spacecraft use small thruster rockets to control their attitude. To rotate a spacecraft, a pair of thruster rockets on opposite sides of the vehicle are fired in opposite directions. To stop the rotation, a second pair is fired to produce an opposing force.

参考译文：飞机构造与控制

飞机和飞船的飞行环境完全不同，所以它们需要不同的方法来引导它们的飞行以及保持方向。

为了提供其飞行的稳定性与控制，大多数飞机使用与机翼相同工作原理的各种控制面，而航天器则利用推力和陀螺力。

飞行员如何控制飞机？飞机机翼和尾翼上的可动表面允许飞行员操纵飞机并控制飞机的姿势和方向。这些控制表面的工作原理与机翼升降原理相同，它们能够在空气产生气压差的时候，在所需的方向上对飞机产生力的作用。

机尾的主要作用是提供稳定性。一架稳定的飞机往往会在被风吹倾斜后立即恢复正常，就像位于碗底的球在受到干扰后会回到碗底中心。

什么是飞机的"尾鳍"？垂直稳定器，或称尾鳍，能够使飞机在运动方向保持平衡。当飞机直线移动时，空气以相等的力压在其两个表面上。但如果飞机向右或向左转动，则稳定器一侧的空气压力增加，另一侧则降低。这种不平衡的压力使尾翼调整回原来路线。

什么是飞机的"后翼"？像垂直稳定器一样，水平稳定器有助于使飞机与其运动方向保持一致。如果飞机向上或向下倾斜，空气压力在稳定器的一侧增加，另一侧降低，并将其推回到其原始位置。稳定器能保持机尾向下，这是飞机的重心始终在机翼升力中心的前方的结果。

控制飞机与控制汽车或船有何不同？对于可以在三维空间中自由移动的飞机，稳定性和控制性要比汽车或船复杂得多。三维空间的三种运动中的任一种变化都会影响另外两种运动。

想象一下，三条线穿过飞机，在飞机的重心处以直角相交。

(1) 围绕前后轴的旋转称为翻转。

(2) 围绕侧面轴的旋转称为倾斜。

(3) 围绕垂直轴的旋转称为偏航。

1. 副翼控制翻转

在每个机翼的外后缘处，两个副翼分别向上下两个相反的方向移动，减少一个机翼的升力，同时增加了另一个机翼的升力，这使飞机向左或向右翻滚。为了转动飞机，飞行员使用副翼将机翼倾斜到所需的方向。

2. 升降舵控制倾斜

在尾部水平表面上，升降舵向上或向下倾斜，减小或增加尾部的升力。这样会使飞机上升或下降。

3. 方向舵控制偏航

在垂直尾翼上，方向舵从一侧向另一侧旋转，向左或向右推动尾部。飞行员通常使用方向舵和副翼一起转动飞机。

如何在空间控制位姿？因为飞机使用的控制面不能在真空环境中工作，所以航天器需要借助不同形式对其姿态进行控制。为了改变方向，航天器通过为小火箭点火或旋转内轮来施加扭矩（扭转力）。

一些卫星使用反作用轮 ——陀螺仪系统来改变其在太空中的姿势。快速旋转的轮子具有很大的旋转动量，使得航天器可以利用它来改变自己的方向。为了能在任意三维空间中做到这一点，航天器必须使用三个相互垂直的旋转轮。

推力如何帮助飞行器改变姿势？大多数航天器使用小型火箭推进器来控制它们的姿势。为了使航天器旋转，在航天器的相对侧上发射反方向的火箭推进器。为了停止旋转，发射第二组火箭推进器以产生相反的力。

New Words and Expressions:

orientation [ˌɔːriˈenˈteɪʃ(ə)n; ˌɒr-] n.方向；定向；适应；情况介绍；向东方

maneuver [məˈnuːvə] n.[军] 机动；演习；策略；调遣

orientation [ˌɔːriˈenˈteɪʃ(ə)n] n.方向；定向；适应；情况介绍；向东方

tail fin 鱼的尾鳍；直尾翼，垂直尾翼

pivot['pɪvət]n.枢轴；中心点；旋转运动

rear wing 后翼子板

horizontal stabilizer[电子] 水平稳定器

yaw [jɔ:]n.（火箭、飞机、宇宙飞船等）偏航

pitch [pɪtʃ] vi.倾斜；投掷；搭帐篷；坠落；俯仰

dimension [dɪ'menʃ(ə)n] n.方面;[数] 维；尺寸；次元；容积 vt. 标出尺寸

vertical axis[力] 垂直轴；纵轴

elevator ['elɪveɪtə] n.电梯；升降机；升降舵；起卸机

rudder ['rʌdə] n.船舵；飞机方向舵

thrusters ['θrʌstə] n.（火箭）[航] 推进器；向上钻营的人

spinning ['spɪnɪŋ] n.纺纱

torque [tɔ:k] n.转矩，[力] 扭矩；项圈；金属领圈

dimensions[dɪ'menʃənz] n. 规模；大小

Questions:

Answer the following questions according to the text.

（1）How does a pilot control an airplane?

（2）How do you control attitude in space?

（3）What is an airplane's "tail fin" for?

Passage Two: The Difference Between Airbus and Boeing Aircraft Design

Large Boeing aircraft have an inboard aileron/high speed aileron/flaperon (marked dark grey in the Fig.4-1 below) near the wing kink, large Airbus aircraft without.This flaperon on Boeing aircraft is usually used to roll at high speeds, because the efficiency of the "normal" ailerons is decreased due to high dynamic pressure and wing flexibility.

However, Airbus aircraft use "normal" ailerons to roll at every speed. On one hand, they save some weight for not having additional systems for the flaperon. On the other hand, to maintain the aileron efficiency at high speeds, the wing has to be stiffer, which increases the weight.

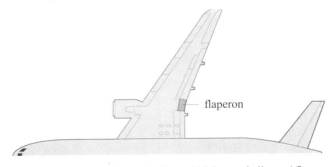

Fig.4-1　Position of inboard aileron / high speed aileron / flaperon

Boeing 737 is a smallest of Boeing aircraft on sale today as shown in Fig.4-2. If you want to differentiate Boeing 737 from other aircraft, it's pretty easy.737 have an extra "wind-breaker" in front of their vertical stabilizer.They have a very pointy nose and a cockpit glass that is aggressive.737's engines are flat on the bottom because of the design legacy. In short, 737 suffer from low ground clearance thus the engines cannot be full-round.737 have quite an expressive winglet. More aggressive than the A320. If we take a look at the new 737-MAX, the aircraft has an "additional" winglet under the existing one, and the engine nacelles have this rigid shape for more efficiency.

Fig.4-2　Boeing 733 aircraft

The 747 was as distinguishable as it is shown in Fig.4-3. It is a large, 4-engined wide-bodied aircraft that have a hump on the front.The newer 747-8 have similar engine nacelles as the 737-MAX and it has no winglet. It now has a more efficient raked wingtip like the 787.The older 747-400 however does have a conventional type winglet, and it is shorter in length than the 737.

Fig.4-3 Boeing 747 aircraft

The A320 (in Fig.4-4) family is the smallest of all airbus commercial jets. It is a direct competitor to the 737 so it have slightly the same size.The A320 have higher ground clearance than the 737.The nose is rounded instead of pointy, and it have more friendly-looking cockpit glass.A320 used to have gated winglet, but newer models now adopt what Airbus called "sharklet" which is basically similar to 737's winglet.

Fig. 4-4 Airbus 320 aircraft

A350 (in Fig.4-5) is the newest aircraft for Airbus that have a very distinctive. Because if visually compared to other Airbus it stands out like an outlier as the design took a bit of a fresh look. A350 have more pitot tube in the nose.It has a "spectacle" on the cockpit window.It has a curved wingtip.

Fig.4-5 Airbus 350 aircraft

More and more, the jets are looking pretty much alike, which isn't too surprising given that some of the best aerodynamicists in the world all are focused on coming up with fuselage shapes and combinations of flight control surfaces that will be as reliable and efficient as possible. The ideal combinations of those features are pretty likely to bear a strong resemblance to each other.

Airbuses are optimized for certain types of operations, and Airbus pilots tend to love them for the efficiency of their automation. On the other hand, Boeings tend to be pretty robust designs that aren't necessarily as fuel-efficient, but are much loved by their pilots for the transparency their flight management systems offered.

The only potentially noteworthy difference I've perceived over the past couple of years is an apparent uptick in instances where Airbus pilots got into trouble because the automation was confusing them, and basic airmanship seemed lacking in trying to sort things out. This isn't necessarily the fault of the aircraft or the manufacturer, but such occurrences haven't seemed to happen in Boeings.

Other differences include engine placement, where Airbus generally places the engines under the wings, while Boeing mounts them on the forward of the wing. However, this placement changes depending on the design of the planes. One major difference is the electronics that is used in the planes. Airbus planes are controlled using sidestick, this is similar to a joystick that is used to play computer games. Boeing uses a central yoke and has a steering wheel design that is used as the main controller. Airbus also has incorporated hard protection using the fly by wire systems. This means that system of the planes cannot be overrun. The pilot cannot assume full control of the system and do something such as stall it. In Boeing, the pilot can perform all these functions by simply overriding the pilot system. The controls in Boeing also have feedback or feels, which are missing in Airbus controls.

参考译文：空中客车与波音飞机设计的主要区别

大型波音飞机在翼扭结附近有一个内侧副翼/高速副翼/襟副翼（见图4-1中标记为深灰色部分），大型空客飞机则没有。波音飞机上的这种襟副翼通常适用于飞机高速时的滚动状态，因为"正常"副翼的效率会由于高速压和机翼柔性而降低。

　　然而，空中客机使用"正常"副翼以正常速度滚动。一方面，它们节省了一些重量，因为没有额外的襟副翼系统。另一方面，为了在高速下保持副翼效率，机翼必须更硬，这样又增加了重量，从而保持平衡。

　　波音737是至今生产的波音飞机中体积最小的一种（见图4-2）。如果你想区别波音737与其他飞机，是十分容易的。737在其垂直稳定器前面有一个额外的"通风孔"。他们有一个非常尖的前凸并且驾驶舱玻璃咄咄逼人。由于设计上还存在遗留系统，737的发动机扁平地置于底部。简而言之，737底部与地面间隙很小，因此发动机不具全能性。737的小翼极具表现力，比A320更具侵略性。如果我们来看看新的737-MAX，那么这架飞机在现有的小翼下有一个"额外的"小翼，发动机舱具有这种刚性的形状，以便提高效率。

　　从图4-3可以看出，747也是容易区分的。它是一台大型的4引擎宽体飞机，前面有一个隆起。与新的747-8相比，它具有与737-MAX类似的发动机舱，并且没有小翼。现在的747有一个像787这样更有效率的前锋翼尖。旧的747-400确实有传统的小翼，长度比737短。

　　A320系列是所有空客商用飞机中最小的（见图4-4）。这是737的直接竞争对手，因此它们的尺寸大致相同。A320的离地净高比737大。前凸圆润而不是尖头，它有看上去更友善的驾驶舱玻璃。A320曾经有过门襟翼，但较新的机型现在均采用了空中客车公司所说的"小鲨鱼"结构，基本上与737的小翼相似。

　　A350（见图4-5）是空中客车最新的一款飞机，是非常有特色的。因为如果与其他空中客车的飞机在视觉上相比，它在设计方面用了一些新意，它就像一个另类。A350在前凸上有更多的皮托管，它在驾驶舱窗户上有一个"视景"，并且有一个弯曲的翼尖。

　　越来越多的喷气式飞机看起来非常相似，这并不令人惊讶。因为世界上最好的空气动力学家都专注于提出机身形状和飞行控制表面结合的提议，这将使之变得可靠和高效。这些特征的理性结合本身就具有很强的相似性。

　　空中客车针对某些类型的操作进行了优化，空中客车公司的飞行员会更喜欢这种自动化操作的有效性。反观，波音公司在这方面则不同，他们更倾向于稳健的设计。这种设计不一定能提高燃料利用率，但是他们的飞行员非常喜欢他们的飞行管理系统提供的透明操作性。

　　在过去几年中，唯一潜在的、值得注意的便是因为自动化操作使空客飞行员感到困惑，由于飞行员似乎缺乏排除故障的经历及能力，他们陷入困境的情况明显上升。这不一定是飞机或制造商的过错，但波音公司似乎并没有发生过这种情况。

其他不同之处还包括发动机配置，空中客车通常将发动机放置在机翼下方，而波音将其安装在机翼的前方。然而，这个位置会根据飞机的设计而改变。一个最主要的区别是飞机上使用的电子设备。空中客车飞机使用侧杆进行控制，这与用于玩电脑游戏的操纵杆相似。波音使用中央轭架，并设计具有用作主控制器的方向盘。空中客车公司还通过使用飞行线控系统加强硬保护度。这意味着飞机的系统不能被飞行员进行例如停顿的完全操控。在波音方面，飞行员可以通过简单地定义飞行系统来执行这些所有功能。波音公司的控制系统具有反馈信息的功能，这在空中客车的控制系统中是不具备的。

New Words and Expressions:

inboard ['ɪnbɔːd] *adj.* 内侧的；舱内的

aileron ['eɪlərɒn] *n.* 副翼

flaperon['flæpə,ran] *n.* [航] 襟副翼

wing kink 翼扭结

dynamic [daɪ'næmɪk] *adj.*动态的；动力的；动力学的；有活力的

vertical stabilizer 垂直尾翼

pointy ['pɒɪntɪ] *adj.* 尖的；非常尖的

cockpit ['kɒkpɪt] *n.*驾驶员座舱；战场

legacy ['legəsɪ] *n.*遗赠；遗产

clearance ['klɪər(ə)ns] *n.*清除；空隙

nacelle [nə'sel] *n.*气球吊篮；飞机的驾驶员室；飞机的引擎机舱

hump [hʌmp] *n.*驼峰；驼背；圆形隆起物

wingtip['wɪŋ,tɪp] *n.*（飞机的）翼尖

pitot tube[流] 皮托管（流速计）；皮托静压管

fuselage ['fjuːzəlɑːʒ; -lɪdʒ] *n.* [航] 机身（飞机）

automation [ɔːtə'meɪʃ(ə)n] *n.*自动化；自动操作

transparency[træn'spær(ə)nsɪ] *n.*透明，透明度；幻灯片；有图案的玻璃

electronic [ɪ,lek'trɒnɪk] *adj.*电子的

airmanship ['ɛrmənʃɪp] *n.*飞行术；导航技术

placement ['pleɪsmənt] *n.*布置；定位球；人员配置

controller [kən'trəʊlə] *n.*控制器；管理员；主机长

Questions:

Answer the following questions according to the text.

（1）What is the characteristic of Boeing 747?

（2）Which aircraft has a flaperon?

（3）Which aircraft have humps in the Boeing series?

Unit Five

Passage One: Drivers and Challenges for Aerospace Manufacturing

Manufacturing is a major driver for development in the aerospace industry, with productivity increases having a dramatic effect on the number of jobs that exist. The high average age of workers within the aerospace industry presents more additional challenges. Currently, the average age of the engineers who support manufacturing work is 54 years, and the average of the blue-collar team that actually put the product together is 51 years. The imminent retirement of many of these engineers and blue collar workers, combined with the inability to hire new workers in the current economic climate, presents a tremendous challenge of transferring knowledge and skills.

In 2005, the U.S. aerospace industry constituted 82 percent of the global aerospace market. However, by 2015, this had decreased to 62 percent. Although the United States has dominated the global aerospace industry in the past, the rest of the world is catching up as they realize that aerospace industry jobs are key to their own economies. Jobs in the aerospace industry are desirable because they are high paying and high technology. During the development of the Boeing 777, approximately 700,000 people worldwide were involved in some way, for example, through supply chain networks. The commercial transports of manufacture therefore has an enormous impact on the global economy.

Boeing is the largest exporter within the United States, and the U.S. aerospace industry is the single largest contributor to the nation's positive surplus of trade. However, between 2000 and 2010 there was a drop in the surplus of trade by almost 50 percent as a result of foreign competition. The product that drives the single largest portion of the trade surplus is commercial transport planes. During the past 30 years, passenger growth has been constant and the demand for commercial transports has grown. However, over the past few years, there has been a significant decline in passenger traffic in the United States. Several U.S. airlines have declared bankruptcy as a result. In Europe and the rest of the world, however, passenger growth has regained the place that it held prior to September 11, 2011.

In the long term, there is a huge market for the aerospace industry. Currently, about 10,000 commercial transports operate worldwide. With an estimated modest growth rate of 3.7 percent per year in passenger and freight travel, it is estimated that a total of 4,500 replacements will be required by 2019 and a total of 9,000 replacements will be required by 2029. Over the next 20 years, this market for commercial transports alone is predicted to be worth about $1.8 trillion.

To keep pace with the competition, Boeing faces several manufacturing process challenges. The order and delivery cycles for transport aircraft since 1958 are fairly regular, between 10 and 12 years long. During periods when the airline industry reaps profits, orders for new transport aircraft are placed. However, due to the length of time required for manufacturing, by the time the planes are ready, the airline companies are often facing a low market and therefore withdraw their orders. The ability of Boeing to shorten the manufacturing and design build time is a critical challenge. Manufacturing optimization goals at Boeing include a reduction in cycle time by 60 percent and a reduction in manufacturing hours by 50 percent.

The aerospace industry is a very mature industry, and Boeing has been building airplanes for almost 100 years. Technology solutions within this industry are therefore evolutionary rather than revolutionary when compared with other fields such as biotechnology, where technology is exploding. However, some areas of manufacturing technology that are a high priority for improvement include:

（1）Single source of production data;

（2）Integrated design/build/quality and supplier processes;

（3）Simplified manufacturing planning;

（4）Design for manufacturing.

Boeing has already made progress in a number of these areas, including the integration of design and manufacturing. Several computer-based models are currently available that are used by engineers to design products. Duplication is avoided and efficiency significantly increased when the same computer-based model is used on the factory floor. In addition, progress has been made in validating systems through electronic simulation. Today, virtual manufacturing is used to virtually demonstrate the prototype instead. This results in both time and cost savings.

Production system from the early 1990s, are now being used by the aerospace industry to eliminate waste, eliminate unnecessary inventory. Boeing is currently spending a lot of time on lean activities. One example is in the production of composites. Certain pieces of large and capital-intensive equipment, such as autoclaves and nondestructive testing equipment, have been completely eliminated by the use of new materials approaches.

On planes such as the 737 and 747, the aerospace industry has shifted to the use of moving production lines in order to shorten assembly cycle times. This requires modification of all of the systems that support the manufacturing activity, which in turn has resulted in a rethinking of the entire production system. For example, computer modeling and laser trackers have replaced large fixtures and tools for assembly.

Changing the part count and the design process can have a huge impact on manufacturing and manufacturing costs. There is currently a debate at Boeing over new ways of thinking about common parts. For example, a strut is a part that holds the engine onto a commercial transport. It consists of hundreds of pieces. New ways of using casting technology can reduce this number of pieces by 90 percent and can reduce the number of fasteners needed by 80 percent. This will improve the durability of the product.

Boeing's near-term manufacturing priorities are reducing and standardizing holes and fasteners; improving composite processing; shortening assembly and cycle times; monolithic structures; turning suppliers into partners; and integrating design and manufacturing. In addition to these near-term priorities, Boeing has identified a number of long-term goals, including simplifying joining techniques; introducing advanced composites by eliminating autoclaves and inspections; reducing the part count; and integrating the production system. During the past 15 years, the U.S. manufacturing industry has nearly doubled its spending on manufacturing research and development, while federal funding has decreased. Industry funds, however, are focused on very near term needs. Federal funding is needed to focus on the high-risk activities that may have alarge impact on the industry in the long term.

参考译文：航空制造业的动力与挑战

制造业是航空航天业发展的主要驱动因素，生产力的提高对现有就业人数有显著的影响。航空航天业工人的平均工作年龄高，这带来了更多的附加挑战。目前，从事制造工作的工程师平均年龄为54岁，实际将产品整合在一起的蓝领队伍的平均年龄为51岁。这其中许多工程师和蓝领工人即将退休，加上在当前经济环境下很难招聘新员工，这是对转移知识和技能的巨大挑战。

2005年，美国航空航天业占全球航空航天市场的82%。然而，到2015年，这已经下降到了62%。虽然美国过去一直主宰着全球航空航天业，但世界其他地区正在追赶，因为他们意识到航空航天工业是本国经济的关键。航空航天业是高薪和高科技工作的理想之选。在波音777的发展过程中，全球约有70万人以某种方式参与，例如通过供应链网络。因此制造业的商业运转对全球经济有着巨大的影响。

波音是美国最大的出口商，美国航空航天业是美国贸易顺差的最大贡献者。然而，在2000—2010年间，由于国外竞争，贸易顺差下降了近50%。改变贸易顺差最大的产品是商业运输机。在过去30年里，客运增长一直不变，对商业运输的需求也在增长。不过，近几年来，美国的客运量大幅度下降。几家美国航空公司最终宣布破产。然而,在欧洲和世界其他地区,客运量的增长已恢复了它在 2011 年 9 月11日之前的位置。

从长远看，航空航天业有着巨大的市场。目前，约10 000个商业运输机业务遍布全球。预计乘客和货运旅游年均增长率为3.7%，估计到2019年将需要4 500个替代品，到2029年将需要9 000个替代品。在接下来的20年中，单靠商业运输市场预计价值约为1.8万亿美元。

为与竞争对手保持同步，波音公司在制造过程中面临着几个挑战。1958年以来，运输机的订单和交付周期相当规律，在10～12年之间。在航空业获利的期间，又有新的运输机订单。然而，由于制造所需的时间长，飞机准备就绪时，航空公司往往面临市场低迷，因此撤销订单。这对波音缩短设计和制造的周期是一个关键的挑战。波音公司的制造优化目标包括将全周期时间缩短60%，制造时间缩短50%。

航空航天业是一个非常成熟的行业，波音公司已经制造飞机近百年。与其他技术爆炸的领域（例如生物技术领域）相比，这个行业的技术解决方案是进化性的而不是革命性的。然而，一些亟待改进的制造技术领域包括：

（1）生产数据的单一来源；

（2）集成设计/制造/质量和供应商流程；

（3）简化制造规划；

（4）制造设计。

波音在一些领域已经取得进展，包括设计和制造的整合。目前有几种基于计算机的模型可供工程师用于产品设计。当在工厂车间使用基于计算机的模型时，可以避免重复，显著提高效率。此外，通过电子仿真验证系统取得一定进展。今天，虚拟制造用于虚拟展示原型，这样可以节省时间和成本。

源于1990年代初的制造方法现在正被用于航空航天业，以消除浪费，消除不必要的库存。波音公司目前在花费大量的时间以达到精益求精。一个例子是复合材料的生产。一些大型和资本密集的设备，如高压釜和非破坏性测试设备，已经通过使用新材料的方法完全消除。

在737和747等飞机上，航空航天业已经转向使用移动式生产线，以缩短装配周期。这需要修改支持制造活动的所有系统，这反过来又导致对整个生产系统的反思。例如，计算机建模和激光跟踪器已经替换大型固定装置和工具进行装配。

改变零件数量和设计过程可能对制造和制造成本产生巨大的影响。目前波音公司正在就共同的部分思考新的方法进行辩论。例如，支柱是将发动机保持在商业运输机上的部分，它包括数百个部件。使用铸造技术的新方法可以将这一数量减少90%，可以减少80%的紧固件数量。这将提高产品的耐久性。

波音公司近期制造的着重点是减少钻孔数目和规范紧固件的标准；改善复合加工；缩短组装和循环时间；整体结构；将供应商转为合作伙伴；并整合设计与制造。除了这些近期的优先事项外，波音还确定了一些长期目标，包括简化连接技术；引进先进复合材料，消除高压釜和检测；减少零件数；以及整合生产系统。在过去15年中，在联邦支持资金减少情况下美国制造业的研发支出几乎翻了一番。然而，行业基金则专注于短期需求。联邦资助需要重点关注长期可能对该行业产生重大影响的高风险活动。

New Words and Expressions:

aerospace ['eərəspeɪs]*n.* 航空宇宙；[航] 航空航天空间

surge [s3:dʒ]*n.* 汹涌；大浪，波涛；汹涌澎湃；巨涌

economic [ˌi:kə'nɒmɪk; ek-]*adj.* 经济的，经济上的；经济学的

imminent ['ɪmɪnənt]*adj.* 即将来临的；迫近的

tremendous [trɪ'mendəs]*adj.* 极大的，巨大的；惊人的；极好的

dominated ['dɒmineitid]*adj.* 占主导地位的；强势的；占统治地位的

desirable [dɪ'zaɪərəb(ə)l]*adj.* 令人满意的；值得要的

domestically [də'mɛstɪkli]*adv.* 国内地；家庭式地；适合国内地

surplus ['s3:pləs]*n.* 剩余；[贸易] 顺差；盈余；过剩

commercial [kə'm3:ʃ(ə)l]*adj.* 商业的；营利的；靠广告收入的

bankruptcy ['bæŋkrʌptsɪ]*n.* 破产

boeing ['bəuiŋ]*n.* 波音（客机）；波音公司

reaps [ri:p]*v.* 收割；收获；获得

optimization [ˌɒptɪmaɪ'zeɪʃən]*n.* 最佳化，最优化

inspectors [ɪn'spɛktər] *n.* [经管] 检查员，巡视员；检查器

demonstrate ['demənstreɪt]*vt.* 证明；展示；论证

Questions:

Answer the following questions according to the text.

（1）What are the main drivers of employment in the aerospace industry?

（2）What is the largest aircraft export company in the United States?

（3）Summarize the challenges of aviation manufacturing.

Passage Two: Design Technology of Aircraft Structural Parts Based on Manufacturing

The full three-dimensional design technology based on the Macro Block Design

（MBD） has been gradually applied to change the aircraft product development methods to shorten the development cycle of aircraft products in the domestic aircraft design process. However, the design process of aircraft structural part is focusing on functionality and lacking of consideration of its manufacturing process which leading to the design process and manufacturing process fracture and increasing the difficulty of manufacturing and production costs. Moreover, aircraft structural parts is a typical production mode of small batch and multi variety which is difficult to accumulate and reuse the knowledge by a large number of production mode. Therefore, the establishment of aircraft structural part design and manufacturing knowledge acquisition, reuse and integration technology, in order to break the information barrier between design and manufacture. We need to consider the manufacturability in the design process of aircraft structure parts to realization manufacturing oriented intelligent design. The core of manufacturing oriented design lies on knowledge acquisition, expression and modeling technology.

The processing technology of aircraft structural parts is complicated. How to acquire and express the machining process knowledge of aircraft structural parts accurately and completely to promote the intelligence of design process for aircraft structural parts is important. Although the machining process knowledge of aircraft structural parts is complex and various. But the vast number of process knowledge is related to the features of the parts. Feature technology plays an important role in information integration as a means of digital manufacturing. In the application of traditional research and commercial software, feature technology has been widely used in the field of CAD/CAM. Especially it has been widely accepted in the field of CAM, which being considered that can effectively improve the process of programming efficiency, conducive to the integration and sharing of knowledge and improving the manufacturing efficiency and quality of the parts and so on. But how to make the feature technology as the link between the design process and manufacturing process is worth discussing and focusing on. Feature is a group geometries with specific attributes and interrelation which including a general description of the parts shape, process and function information. It is the best carrier to integrate information of design and manufacturing which can effectively improve product design automation. Features can transfer information to the downstream manufacturing process and communicate design process and the manufacturing process as the link. Through the establishment of knowledge and model based on the feature, we can build the features knowledge base for the design process to achieve sharing knowledge in the design and manufacturing process, which provides the basis for manufacturing oriented intelligent design.

The aircraft is a variety of high-tech integrated body which representing the level of a national industrial design and manufacturing and assembly. With the slowdown of the global economy, the development of manufacturing industry into a difficult period. The enterprises need to change the traditional mode that focused on the production task and do not pay attention to the production cost control. Especially in aircraft structural parts, the material cost is high. And the machining process is long. Unreasonable design is very easy to cause the quality of the products is difficult to control, which increasing the difficulty of product manufacturing and production costs. Therefore, it need to build the fast, high-quality design methods for aircraft structural parts urgently to enhance the core competitiveness of the aircraft manufacturing industry. The feature types of aircraft structural parts can be divided into slot feature, rib feature, hole feature and contour feature, which we define as geometry features. Aircraft structural parts not only need to define the part geometry in the design process. But also need to define parameter information including the technology of parts and material and precision and so on in the design process, which we will define as process features, it is an important basis for the machining and inspection. In the design process of aircraft structure parts, the feature based parametric modeling technology and the process knowledge interpreter based on feature is application in the building of the parts' geometric and the process feature, which making machining process knowledge integrated into the aircraft structure parts design fully. The design process become one part of the manufacturing process and it realize the intelligent design of aircraft structural parts.

Modern aircraft research and development are general using of multi-manufacturer and remote collaborative development model. The production mode of small batch and multi variety of aircraft structural parts. Especially in the development and production process is accompanied by a large number of design changes which required manufacturing process to respond quickly, including manufacturer selection, process and resource preparation and so on. At the same time, in the traditional aircraft structure development process, the design process and manufacturing process is relatively independent, which only has the design to manufacture data transmission in one-way. The manufacturing process data can not be feedback to design apartment. Therefore, to establish the multi-information communication channels of the aircraft structural parts' design and manufacturing process based on digital technology and construction the collaborative platform of design and manufacturing for aircraft structural part achieve the aircraft structure life cycle data sharing of multi sectoral and area, in order to truly realization collaborative design and collaborative manufacturing and design and manufacture collaboration.

From the overall objective of collaborative research and development, the product digital collaborative development has two models. It is the horizontal respectively. Airbus applied collaborative technology research in Viva Ce system research project from feasibility study and concept design until the detailed design of the entire life cycle, and constructs a multidisciplinary collaborative development system framework of MDO. It applied in three aviation of helicopter, aircraft and engine. In the process of the Airbus A380 development, the idea of multi-subject cooperative development is fully utilized in the design of the aircraft. But these digital collaborative development methods are just to achieve the data from upstream to downstream in one-way or horizontal transmission between the same level and lack of downstream data to the upstream feedback. The design and manufacturing collaborative platform for aircraft structure part of the development of big data technology and network technology provides the basic conditions for construction of the collaborative platform. In the collaborative platform, the software resources (CAD, CAE software), hardware resources (numerical control equipment, testing equipment, etc.) and knowledge resources (process knowledge database, parameter database) that belong to different levels, different regions and design departments and different regional manufacturer can achieve horizontal and vertical sharing.

The direct transmission of intelligent coordination and scheduling data of the design and manufacture process can complete aircraft structure design and manufacturing process fast to achieve synchronization of distributed resources. At the same time, aircraft structural parts manufacturing process data can rapidly and timely feedback to the design department by collaborative platform to guidance and optimization the design process of aircraft structural parts which could enhance aircraft structure parts manufacturability and reduce the aircraft structure parts and aircraft product development cycle.

参考译文：基于制造的飞机结构件设计技术

　　基于宏数据块设计（MBD）的全三维设计技术已逐渐应用于改变飞机产品开发方式，缩短飞机产品在国内飞机设计过程中的开发周期。然而，飞机结构部件的设计重点集中在功能上，缺乏对制造过程的考虑，导致设计过程和制造过程分离，增加制造难度和生产成本。此外，飞机结构部件是典型的小批量生产方式，难以通过大量生产的方式积累和重用知识。因此，获取飞机结构部件设计和制造信息，再利用整合技术，以打破设计与制造之间的信息障碍。我们需要考虑飞机结构部件设计过程中的可制造性，以实现以制造为导向的智能设计。制造导向型设计的核心在于知识的获取与表达以及建模技术。

　　飞机结构部件的加工技术复杂，如何准确、完整地获取和表达飞机结构部件的加工过程知识，以促进飞机结构部件设计过程的智能化很重要。飞机结构部件的加工过程知识虽然复杂多样，但是大量的流程知识是与零件的特征相关的。特征技术在信息集成中作为数字制造的手段发挥着重要作用。在传统研究和商业软件的应用中，特征技术已被广泛应用于CAD／CAM领域。特别是在CAM领域，特征技术被广泛使用，被认为可以有效地提高编程效率，有利于知识的整合和共享，提高零件的制造效率和质量等。但是，如何使特征技术成为设计流程与制造流程之间的纽带，值得关注和讨论。特征是具有特定属性和相关性的几何组，是工件形状，加工过程和功能信息的一般描述。它是集设计和制造信息的最佳载体，可有效提高产品设计自动化。特征可以将信息传递到下游制造过程，并将设计过程和制造过程作为链接。通过基于特征的模型信息的建立，我们可以构建设计过程的特征知识库，在设计和制造过程中实现知识共享，为以制造为导向的智能设计提供依据。

　　飞机是各种高科技的综合体，代表了国家工业设计制造和装配水平。随着全球经济放缓，制造业发展陷入困境。企业需要改变着重于生产任务并且不注意生产成本控制的传统模式。特别是在飞机结构件制造中，材料成本高，加工过程长。不合理的设计很容易造成产品质量难以控制，容易增加工件制造难度和生产成本。因此，要加快建立飞机结构部件快速、高质量的设计方法，这样才能提升飞机制造业的核心竞争力。飞机结构部件的特征类型可以分为槽特征、肋特征、孔特征和轮廓特征，我们将其定义为几何特征。飞机结构件不仅需要在设计过程中定义零件几何，而且还需要在设计过程中定义参数信息，包括部件和材料的技术以及精度等，我们将其定义为过程特征，是加工和检验的重要依据。在飞机结构部分的设计过程中，基于特征的参数建模技术和基于过程知识特征的结合，应用于构建零件几何特征和加工特征，使加工过程知识完全融入飞机结构部件设计中。设计过程成为制造过程的一部分，实现飞机结构部件的智能化设计。

现代飞机的研究与开发采用通用的多制造商和远程协同开发模式。该模式为小批量、多种飞机结构件类型生产模式。特别是在开发和生产过程中伴随着大量设计变更，需要制造过程快速响应，包括制造商选择、流程和资源准备等。同时，在传统的飞机结构开发过程中，设计过程和制造过程相对独立，只能从设计到制造进行单向数据传输，制造过程数据不能反馈给设计部门。因此，基于数字技术，建立飞机结构部件设计制造过程的多信息通信渠道，构建飞机结构部件设计制造协同平台，实现了多部门和区域的飞机结构件生命周期数据共享，以真正实现协同设计和协同制造以及设计和制造协作。

从协同研发的总体目标来看，产品数字协同开发有两个平行的模型。空中客车公司Viva Ce系统中应用协同研究技术，从可行性研究和概念设计到整个详细设计生命周期，构建MDO多学科协同发展体系框架，它适用于直升机、飞机和发动机三种飞行器制造。在空中客车A380开发过程中，多主题合作开发的理念在飞机设计中得到充分利用。但是这些数字协同开发方法只是从上游到下游单向传输或在同一级别间的水平传输，缺乏下游到上游的数据反馈。飞机结构部分的设计制造协同平台大数据技术和网络技术的发展为协同平台的构建提供了基础条件。在协同平台上，属于不同层次、不同地区设计部门的软件资源（CAD、CAE软件），硬件资源（数控设备、测试设备等）和知识资源（过程知识数据库，参数数据库）可以在不同的制造商之间横向和纵向共享。

直接传输智能协调和调度数据可以快速完成飞机结构设计和制造过程，实现分布式资源的同步。同时，飞机结构件制造过程数据可以通过协同平台快速及时地反馈给设计部门，指导和优化飞机结构部件的设计过程，从而提高飞机结构部件的可制造性，减少飞机结构部件和飞机产品开发周期。

New Words and Expressions:

MBD *abbr.*宏程式块设计(macroblock design)

Shorten ['ʃɔːt(ə)n] *vt.* 缩短；减少；变短

Acquisition [ˌækwɪ'zɪʃ(ə)n] *n.* 获得物，获得；收购

Integration [ɪntɪ'greɪʃ(ə)n] *n.* 集成；综合

Oriented ['ɔːrɪentɪd] *adj.* 导向的；定向的；以……为方向的

geometries [dʒɪ'ɒmɪtrɪ]*n.* 几何图形，几何体；几何尺寸

attributes ['ætrə,bjʊt] *n.*属性

representing [ˌrepri'zentɪŋ] *v.* 代表；表示；表现

enterprises ['ɛntər,praɪz] *n.* [经] 企业

exterior [ɪk'stɪərɪə; ek-] *adj.*外部的；表面的；外在的

parameter [pə'ræmɪtə] *n.* 参数；系数；参量

transmission [trænz'mɪʃ(ə)n; trɑːnz-; -ns-] *n.*传动装置，[机] 变速器；传递

sectoral ['sektərə] *adj.* 部门的；行业的；经济领域的

MDO *abbr.*多学科设计优化；多学科优化；设计混合域示波器

collaborative [kə'læbərətɪv] *adj.* 合作的，协作的

Questions:

Answer the following questions according to the text.

（1）What are the aspects of MBD's all-three-dimensional design techniques?

（2）What are the characteristics of traditional research and commercial software applications?

（3）What is the pattern of modern R & D technology development?

Unit Six

Passage One: The Development Trend of Aircraft Manufacturing

As a typical representative of the equipment manufacturing industry, the aviation manufacturing industry is a strategic high-tech industries which reflecting a country's overall national strength and the overall industrial level. It's also one of the ten major areas of China's manufacturing 2025. High efficiency and high quality is an important condition for design and manufacturing of aircraft structural parts to ensure the performance and progress of aircraft manufacturing. It is not only requires the application of the combination of digital technology to aircraft structure design and manufacturing process simply, but also more emphasis on collaborative of design and manufacturing, namely product goals would be set to a driver, the real-time, decentralized design information and manufacturing information will be collecting, optimization and integration by intelligent control system, to better service in product design and manufacturing processes.

In order to better service in the whole process of design and manufacturing of products. Therefore, how to fuse the design process and manufacturing process of aircraft structural parts and build the design and manufacturing system of aircraft structural parts based on intelligent technology is the urgent demand of the equipment manufacturing industry in our country. For the first time, the fourth industrial revolution (industry 4.0) which is dominated by intelligent manufacturing is put forward since 2011 in the German Hanover Fair, other countries also put forward the corresponding development plan for intelligent manufacturing, such as the United States to revitalize manufacturing strategy, Japanese manufacturing white paper and made in China 2025. Visible intelligent manufacturing is becoming a new round of industrial competition in the high ground. Aircraft structural parts is a typical production mode of small batch and multi-variety, which is not conducive to the accumulation and reuse for knowledge. Therefore, it is of great and practical significance to apply intelligent technology in the design and manufacture process of aircraft structural parts.

The American Lockheed Martin in the JSF project used the product lifecycle management (PLM) software as integration platform and restructured company process by application the digital design and manufacturing management and give full play to the partner's optimal ability. The JSF aircraft design time reduced by 50%, 66% reduction in manufacturing time, assembly tooling reduced 90%, discrete parts is reduced by 50%. The design and manufacturing and maintenance costs are reduced by 50%. Boeing company in the new generation of war space vehicle, through the application of MBD/MBI (model based define and model based operating instructions) reduced assembly time 57%, and realization of the following technical breakthrough: design 3D transition to 3D manufacturing data technology, 3D product and process information transmitted to the manufacturing unit workers, the digital data of manufacturing operations at the scene can be acquisition and feedback, and the management of digital quality data and compliance document. Domestic aviation manufacturing enterprises keep up with the world advanced aeronautical manufacturing technology and development of enterprises, through sustained investment in recent years, has been realization of the digital design of the product and manufacturing process based on the model, the key parts manufacturing has basically achieved the machining process of NC. Domestic advanced aeronautical manufacturing enterprise has initially built digital workshop environment, developed a series of business system and software tools, to achieve the digital control of the manufacturing process. It has been basically close to international aviation manufacturing enterprises in the advanced level, having from digital to the automation and intelligent manufacturing development foundation. But from the technology development trend, the existing aircraft structure parts processing workshop is still in primary stage in integration and flexibility and intelligence which mainly relying on advanced and automation of single equipment to realization the efficient manufacturing of aircraft structural parts.

The application of digital technology in the design and manufacture of aircraft structure parts effectively improve the processing quality and processing efficiency of aircraft structural parts, but aircraft structural parts is a small batch and multi-variety production mode which not only need to enhance the quality and efficiency , more need to enhance the flexibility and automation. Future aircraft structural parts design and manufacture should gradually to the intelligent development, breaking the traditional thought of design and manufacturing process is separation, building aircraft structure parts design and manufacture collaborative platform, realization of flexible automatic produce for aircraft structural parts with high efficiency and high quality. But the intelligent manufacturing needs to be based on highly integrated automation production, the domestic aviation enterprise is still a large gap from this requirement. It is imperative to carry out the related key technology research for aircraft structure parts smart design and manufacturing, and promoting the application of the related technologies and standards, to achieve intelligent manufacturing landing in the aircraft structure design and manufacturing field.

参考译文：飞行器制造发展趋势

作为装备制造业的典型代表，航空制造业是反映国家综合实力和整体工业水平的战略性高新技术产业,也是中国2025年制造业十大领域之一。高效高品质是飞机结构件设计制造的重要条件，以确保飞机制造业性能和发展。数字技术的组合不仅要求飞机结构设计和制造过程的应用简单，而且更加强调设计和制造的协作，即将产品目标设定为驱动程序，分散的设计信息和制造信息将通过智能控制系统进行实时收集、优化和集成，更好地在产品设计和制造过程中服务。

为了在产品的设计和制造过程中更好地服务，因此，如何融合飞机结构件的设计过程和制造过程，构建基于智能技术的飞机结构件设计制造体系，是我国装备制造业的迫切需求。首次以智能制造为主的第四次工业革命（工业4.0）自2011年起在德国汉诺威博览会上被提出，其他国家也提出了相应的智能制造发展规划，如美国振兴制造策略、日本制造白皮书、中国制造2025。可视化智能制造正在成为新一轮的高端竞争产业。飞机结构件是小批量和多品种的典型生产方式，不利于知识的积累和再利用。因此，在飞机结构件的设计和制造过程中应用智能技术具有重要的现实意义。

美国洛克希德马丁公司在JSF项目中使用产品生命周期管理（PLM）软件作为整合平台，通过应用数字设计和制造管理重组公司流程，充分发挥合作伙伴的最佳能力。JSF飞机设计时间缩短了50％，制造时间缩短了66％，装配工具减少了90％，分立部件减少了50％，设计和制造及维护成本降低了50％。波音公司在新一代战争空间车辆中，通过应用MBD／MBI（基于概念的模型和基于操作流程的模型），缩短了57％的装配时间，实现了以下技术突破：3D设计向3D制造数据技术转型，将3D产品和流程信息传输给制造单位的工作人员，现场制造操作的数字数据可以被获取和反馈，并管理数字质量数据和合规文件。国内航空制造企业跟上世界先进的航空制造技术和企业发展，近年来通过持续投资已经实现了数字化产品设计和制造过程，关键部件基于模型制造基本实现了NC的加工工艺。国内先进的航空制造企业初步建成了数字车间环境，开发了一系列业务系统和软件工具，实现了制造过程的数字化控制。从数字化到自动化和智能制造的发展来看，已基本接近国际航空制造企业的先进水平。但从技术发展趋势来看，现有飞机结构件加工车间仍处于初级阶段，主要依靠单机先进自动化实现飞机结构件高效制造。

数字技术在飞机结构部件设计和制造中的应用有效地提高了飞机结构件的加工质量和加工效率，但飞机结构件是一种小批量和多品种的生产方式，不仅需要提高质量和效率，更需要增强灵活性和自动化。未来飞机结构件的设计制造应逐步走向智能化发展，打破传统设计和制造分离的思路，构建飞机结构件设计制造协同平台，高效率和高质量地实现飞机结构部件灵活自动化生产。但智能制造需要以高度一体化的自动化生产为基础，国内航空企业与这一要求仍然存在较大的差距。对飞机结构部件智能设计制造进行相关关键技术研究，推动相关技术和标准的应用，实现飞机结构设计制造领域的智能化势在必行。

New Words and Expressions:

aviation [eɪvɪ'eɪʃ(ə)n]*n*. 航空；飞行术；飞机制造业

strategic [strə'ti:dʒɪk]*adj*. 战略上的，战略的

digital ['dɪdʒɪt(ə)l]*adj*. 数字的；手指的

emphasis ['emfəsɪs]*n*. 重点；强调；加强语气

collaborative [kə'læbəretɪv]*adj.* 合作的，协作的

decentralized [di'sɛntrəl,aɪz]*adj.* 分散的；分散化，分散式；分散管理的

optimization [,ɒptɪmaɪ'zeɪʃən]*n.* 最佳化，最优化

equipment [ɪ'kwɪpm(ə)nt]*n.* 设备，装备；器材

accumulation [əkjuːmjʊ'leɪʃ(ə)n]*n.* 积聚，累积；堆积物

lifecycle [laɪf'saɪkl]*n.* 生活周期；生活过程；生命循环

breakthrough ['breɪkθruː]*n.* 突破；突破性进展

acquisition [,ækwɪ'zɪʃ(ə)n]*n.* 获得物，获得；收购

enterprises ['ɛntər,praɪz]*n.* [经] 企业

aeronautical [,eərə'nɔːtɪkl]*adj.* 航空的；航空学的；飞机驾驶员的

flexibility [,fleksɪ'bɪlɪtɪ]*n.* 灵活性；弹性；适应性；柔性

enhance [ɪn'hɑːns; -hæns; en-]*vt.* 提高；加强；增加

separation [sepə'reɪʃ(ə)n]*n.* 分离，分开；间隔，距离

Questions:

Answer the following questions according to the text.

（1）Can the aviation industry reflect the country's situation?

（2）Overview of the process of intelligent manufacturing.

（3）What are the features of the JSF project?

Passage Two: Advanced Composite Materials

Composite materials consist of a combination of materials that are mixed together to achieve specific structural properties. The individual materials do not dissolve or merge completely in the composite, but they act together as one. Normally, the components can be physically identified as they interface with one another. The properties of the composite material are superior to the properties of the individual materials from which it is constructed.

An advanced composite material is made of a fibrous material embedded in a resin matrix, generally laminated with fibers oriented in alternating directions to give the material strength and stiffness. Fibrous materials are not new; wood is the most common fibrous structural material known to man.

Applications of composites on aircraft include:

（1） Fairings；

（2） Flight control surfaces；

（3） Landing gear doors；

（4） Leading and trailing edge panels on the wing and stabilizer；

（5） Floor beams and floor boards；

（6） Vertical and horizontal stabilizer primary structure on large aircraft；

（7） Primary wing and fuselage structure on new generation large aircraft；

（8） Turbine engine fan blades；

（9） Propellers.

An isotropic material has uniform properties in all directions. The measured properties of an isotropic material are independent of the axis of testing. Metals such as aluminum and titanium are examples of isotropic materials. A fiber is the primary load carrying element of the composite material. The composite material is only strong and stiff in the direction of the fibers.

Unidirectional composites have predominant mechanical properties in one direction and are said to be anisotropic, having mechanical and physical properties that vary with direction relative to natural reference axes inherent in the material. Components made from fibers in forced composites can be designed so that the fiber orientation produces optimum mechanical properties, but they can only approach the true isotropic nature of metals, such as aluminum and titanium.

A matrix supports the fibers and bonds them together in the composite material. The matrix transfers any applied loads to the fibers, keeps the fibers in their position and chosen orientation, gives the composite environmental resistance, and determines the maximum service temperature of a composite.

Strength characteristics structural properties, such as stiffness, dimensional stability, and strength of a composite laminate, depend on the stacking sequence of the plies. The stacking sequence describes the distribution of ply orientations through the laminate thickness. As the number of plies with chosen orientations increases, more stacking sequences are possible. For example, a symmetric eight-ply laminate with four different ply orientations has 24 different stacking sequences.

The strength and stiffness of a composite buildup depends on the orientation sequence of the plies. The practical range of strength and stiffness of carbon fiber extends from values as low as those provided by fiberglass to as high as those provided by titanium. This range of values is determined by the orientation of the plies. Proper selection of ply orientation in advanced composite materials is necessary to provide a structurally efficient design. The part might require $0°$ plies to react to axial loads, $\pm 45°$ plies to react to shear loads, and $90°$ plies to react to side loads. Because the strength design requirements are a function of the applied load direction, ply orientation and ply sequence have to be correct. It is critical during a repair to replace each damaged ply with a ply of the same material and ply orientation.

The fibers in a unidirectional material run in one direction and the strength and stiffness is only in the direction of the fiber. Pre-impregnated (prepreg) tape is an example of a unidirectional ply orientation. The fibers in a bidirectional material run in two directions, typically $90°$ apart. A plain weave fabric is an example of a bidirectional ply orientation. These ply orientations have strength in both directions but not necessarily the same strength. The plies of a quasi-isotropic layup are stacked in a $0°$, $-45°$, $45°$ and $90°$ sequence or in a $0°$, $-60°$ and $60°$ sequence. These types of ply orientation simulate the properties of an isotropic material. Many aerospace composite structures are made of quasi-isotropic materials.

All product forms generally begin with spooled unidirectional raw fibers. An individual fiber is called a filament. The word strand is also used to identify an individual glass fiber. Bundles of filaments are identified as tows, yarns, or rovings. Fiberglass yarns are twisted. Tows and rovings do not have any twist. Most fibers are available as dry fiber that needs to be impregnated (impreg) with a resin before use or prepreg materials where the resin is already applied to the fiber.

In short, advanced composites play a revolutionary role in the manufacturing industry. In the future manufacturing industry will continue to play its characteristics.

参考译文：先进复合材料

复合材料是混合在一起以实现特定结构性质的组合材料。单一材料不会在复合材料中完全溶解或完全合并，而是一起作用。通常，组件可以通过它们彼此之间的接口进行物理识别。复合材料的性能优于其组成的各种单一材料的性能。

先进的复合材料由嵌入在树脂基体中的纤维材料制成，通常纤维通过定向叠层获取材料强度和刚度。纤维材料不是新材料，例如木材是人们已知的最常见的纤维结构材料。

复合材料在飞机上的应用包括：

（1）整流罩；

（2）飞行控制面；

（3）起落架门；

（4）机翼上的前翼和后缘面板；

（5）地板梁和地板；

（6）大型飞机的垂直和水平稳定器初始结构；

（7）新一代大型飞机的主翼和机身结构；

（8）涡轮发动机风扇叶片；

（9）螺旋桨。

各向同性材料在各个方向具有均匀的性质。各向同性材料的测量性质是独立于测试轴。如金属铝、钛是各向同性材料的例子。纤维是复合材料的主要承载部件材料。复合材料只有在纤维方向拥有强度和刚度。

单向复合材料在某方向上具有主要的机械性能，是各向异性的，机械和物理性质相对于材料固有的自然参考轴随方向变化。通过设计增强复合定向纤维材料，产生最佳的机械性能，但它们只能接近如铝和钛等金属的各向同性。

基材支撑纤维并将它们结合在复合材料中。基材将施加的载荷转移到纤维上，将纤维保持在相应位置和选定的方向上，使复合材料达到环境耐用性，并确定复合材料的最高使用温度。

复合叠层板的强度特性，如刚度、尺寸稳定性和强度取决于层叠的堆叠顺序。层叠顺序描述了通过层板厚度的不同层取向的分布。随着取向层数的增加，就会有更多的堆叠次序。例如，具有四个不同层取向的对称八层叠层材料，具有24个不同的堆叠序列。

复合材料的强度和刚度取决于层的取向顺序。碳纤维的强度和刚度的实际范围从由玻璃纤维提供的最低值至由钛合提供的最高值。值的变化范围由叠层的取向决定。在先进复合材料中适当选择叠层取向是必要的，这样可以提供有效的结构设计。该部件可能需要0°叠层对轴向载荷做出响应，±45°叠层对抗剪切载荷做出响应，90°叠层对反作用力负载做出响应。因为强度设计要求是施加在载荷方向的函数，所以叠层的取向和层序列必须是正确的。在修正过程中，用相同材料和取向的叠层替换每个损坏的叠层是至关重要的。

单向材料中的纤维是在一个方向上运行，强度和刚度仅体现在纤维的方向上。预浸渍（预浸料）胶带是单向层取向的实例。双向材料中的纤维在两个方向上运行，通常相差90°。平纹织物是双向层叠方向的实例。这些叠层取向在两个方向上具有强度，但不一定具有相同的强度。准各向同性叠层的层叠以0°，−45°，45°和90°的顺序或0°，−60°和60°的顺序堆叠。这些类型的叠层取向模拟各向同性材料的性质。许多航天复合结构由准各向同性材料制成。

所有的产品形式通常以卷绕的单向原纤维开始。单根纤维称为单纤维。这个词也用于定义单个玻璃纤维。纤维束定义为丝束、纱线或粗纱。玻璃纤维纱线是扭曲在一起的。丝束和粗纱没有任何扭曲。大多数纤维是干纤维，在使用前需用树脂浸渍（浸渍），或已经将树脂预浸在纤维上了。

总之，高级复合材料在制造业中起到革命性的作用。在今后制造业中也会继续发挥其特点。

New Words and Expressions:

laminated ['læmɪneɪtɪd]*adj*. 层压的；层积的；薄板状的

materials [mə'tɪərɪəlz]*n*. [材] 材料；材质

fairings ['feərɪŋ]*n*. [航][船] 整流罩

stabilizer ['steɪbɪlaɪzə]*n*. [助剂] 稳定剂；稳定器；安定装置

turbine ['tɜːbaɪn; -ɪn]*n*. [动力] 涡轮；[动力] 涡轮机

propellers [prə'pɛlər]*n*. [航][船] 螺旋桨；[印刷] 螺旋辊

isotropic [ˌaɪsə(ʊ)'trɒpɪk]*adj*. [物][数] 各向同性的；等方性的

aluminum [ˌaɪsə(ʊ)'trɒpɪk]*adj*. [物][数] 各向同性的；等方性的

titanium [taɪ'teɪnɪəm; tɪ-]*n.* [化学] 钛（金属元素）

predominant [prɪ'dɒmɪnənt]*adj.* 主要的；卓越的；支配的

anisotropic [ˌænaɪsə(ʊ)'trɒpɪk]*adj.* [物] 各向异性的；[物] 非均质的

properties ['prɒpətɪz]*n.* 性能；道具，内容

fiberreinforced ['fibərriːɪnfɔːst] 纤维复合材料；纤维增强复合材料；强化纤维复合材料

orientation [ˌɔːrɪən'teɪʃ(ə)n; ˌɒr-]*n.* 方向；定向；适应；情况介绍；向东方

maximum ['mæksɪməm]*n.* [数] 极大，最大限度；最大量

stiffness ['stɪfnɪs]*n.* 僵硬；坚硬；不自然；顽固

sequence ['siːkw(ə)ns]*n.* [数][计] 序列；顺序；续发事件

unidirectional [ˌjuːnɪdɪ'rekʃ(ə)n(ə)l]*adj.* 单向的；单向性的

Pre-impregnated 预浸渍纸绝缘

Questions:

Answer the following questions according to the text.

（1）What is the combination of composite materials?

（2）How does the fiber work in the composite?

（3）The fiber orientation and fiber form are briefly described in their own language.

Unit Seven

Passage One: Feature-Based Cutting Parameter Optimization for Aircraft Structural Parts

The geometric state of aircraft structural part is changing dynamically during the machining process. Such geometric changes lead to changes of the rigidity of workpiece features. Knowing the feature rigidity during the machining process is very crucial for the selection of right cutting parameters especially for aircraft structural parts with complex structures. In order to address this issue, an interim feature concept is introduced and the interim feature is modeled by using geometric information of final state and machining process information for aircraft structural parts. The interim feature rigidity is evaluated by simplifying the structures of interim feature. By considering the constraints of the interim feature rigidity, machine tool characteristics, and cutting tool performance, the cutting parameters are optimized using a genetic algorithm.

Cutting technology is multidisciplinary and playing an increasingly important role in manufacturing industry. There should be a strong integration of technologies and management using information technologies. Specially, the integration of cutting technology with the computer-aided process planning (CAPP) has brought great benefit for machining. Cutting parameter optimization is one of the most important aspects in CAPP, and essential to numerical control (NC) machining process, as it is related to machining efficiency, machining quality, as well as machining cost. Improper cutting parameters may lead to low machining efficiency, harm to machine tools and cutting tools, and may also scrap workpieces in some situations. At present, most of the recommended cutting parameters are so conservative that the performance of machine tools is exerted at a very low level. Nevertheless, the global cutting parameters may be improper in some local positions as the properties of the parts vary, which can lead to part deformation in low rigidity area. On the other hand, the properties of the parts change in the machining process, therefore the cutting parameters should change as the changes of part properties, especially for aircraft structural parts which are characterized with thin walls, complex shape, and large dimension. In addition to that, aircraft structural parts are also featured with high material remove rate of more than 95%. The material is removed step by step, and the rigidity of the workpiece changes with machining steps. The cutting force what the in-process workpiece can stand also changes with machining steps. The cutting force is related to cutting parameters, and therefore the selection of cutting parameters plays a vital role to guarantee machining quality, to increase productivity, as well as to reduce machining cost. Whereas, the selection of optimal cutting parameters needs to consider many factors such as the rigidity and strength of structures, specification of machine tool capabilities, and cutting tool performance. The challenges of cutting parameter optimization of the in-process workpiece are the definition and expression of workpiece interim state.

Due to the important role of cutting parameters in machining, cutting parameter optimization has been an active research topic for many decades with the raise of cutting technologies including traditional cutting and non-traditional cutting electrical discharge machining and laser machining. Because of the nonlinear dependence of machining process, especially in milling, the problem is very complex. In order to select a set of proper cutting parameters, many approaches have been proposed in the evolution of cutting parameter optimization. At the beginning, most of the cutting parameter selection approaches are based on engineer experiences, but it is difficult to obtain optimal parameters. At present, experiment methods are widely adopted to get cutting parameters for specified machining conditions. For most of the time, the cutting parameters from experiments can be used safely for simple parts. The problem is that the parameters are relatively conservative, and it is time consuming.

Dynamics factors including cutting force and process stability in machining process are very important constraints for cutting parameter optimization. Researchers have made great contributions to the analytical models for cutting force and process stability, and the analytical models for cutting force prediction theory have been widely used in cutting field. The analytical models for cutting force prediction are also applied, and the stability region obtained is also used as constraints for cutting parameter optimization.

In the current trends of optimizing machining process parameters, artificial intelligence (AI) algorithms, genetic algorithms, and meta-heuristic techniques have been widely used. In terms of artificial intelligence algorithms, someone presented a neural network-based approach to optimize cutting parameters. However, optimization by these methods often ends in local minima or fails to converge on an optimum. Most of evolutionary algorithms and meta-heuristic techniques are inspired by nature or animal behaviors such as particle swarm optimization (PSO), ant colony optimization (ACO), artificial bee colony (ABC). Today's technology is developing rapidly, the future there are many advanced technology, let us wait and see.

参考译文：基于特征的飞机结构件加工参数优化

在加工过程中飞机结构件中几何状态动态变化。这样的几何变化导致工件的刚性特性变化。对于具有复杂结构的飞机结构部件了解加工过程中的刚性特征对选择正确的加工参数尤其重要。为了解决这个问题，介绍了临界特征概念和通过定义使用临界特征最终状态的几何信息来构建飞机结构件加工过程模型。通过简化临界特征结构评估临界特征刚度。考虑到临界特征刚性的限制，用遗传算法优化切削刀具性能、切削路径、切削参数等。

切削加工技术是多学科技术，在制造业中有越来越重要的作用。应该利用信息技术整合工艺和管理信息。特别是计算机辅助生产规划（CAPP）能为加工带来极大的帮助。加工参数优化是CAPP最重要的方面之一，由于它与加工效率、加工质量和加工成本相关，对数控（NC）加工过程至关重要。不合适的加工参数可能会导致加工效率低，对机床和刀具造成损伤，也可能在某些情况下造成工件报废。目前，大多数推荐的加工参数是保守的，使机床的性能只能发挥出非常低的水平。然而，全球通用的加工参数在工件结构变化时，对于某些局部区域并不合适，这可能导致零件在低刚度区域变形。另一方面，零件的属性在加工过程中发生变化，因此切割参数应随着零件属性的变化而变化，特别是对于具有薄壁、形状复杂、尺寸大特征的飞机结构部件而言。除此之外，飞机结构件也具有高达95%的材料去除率。材料逐步拆除，工件刚度随着加工过程而变化。工件可以承受的切削力也随着加工过程而变化，切削力与切削参数有关，因此选择切削参数对保证加工质量、提高生产率、降低加工成本起到至关重要的作用。然而，选择最佳切削参数需要考虑许多因素，如结构的刚度和强度，机床功能规格和切削工具性能等。切削参数优化的挑战来自工件临界状态的定义和表达。

由于切削参数在加工中的重要作用，最近几十年，随着加工技术的提高，包括传统切削和非传统电火花加工和激光加工等的发展，切削参数优化得到了积极的研究。由于在铣削中，非线性依赖程度高，因此加工过程复杂，选择一组适当的切削参数，提出了许多切削参数优化的方法。一开始，大部分切削参数选择基于工程师的经验，难以获得最优参数。目前广泛采用的是针对指定加工条件选择基于工化加工参数。在大多数时候，对简单的零件可以安全地使用实验中的切削参数。实验优化的问题是这些参数是相对保守的，而且耗时。

包括切削力和加工过程稳定性在内的动力学因素对于切削参数优化是非常重要的制约因素。研究人员为此做出了巨大贡献，建立切割力和工艺稳定性的分析模型，切削力预测理论的分析模型已被广泛应用于加工领域。切削力预测分析模型和获得的稳定区域也被用作加工参数的优化约束条件。

目前，在切削参数优化趋势中，人工智能（AI）算法、遗传算法和元启发技术已经得到了广泛应用。在人工智能算法方面，有人提出了一种基于神经网络的方法进行切削参数优化。但是，通过这些方法进行优化通常以局部最小值结束或不能收敛于最佳值。大多数进化算法和元启发式技术受到自然或动物行为的启发，例如粒子群优化（PSO）、蚁群优化（ACO）、人造蜂群（ABC）。当今科技正在飞速发展，未来还有很多先进的技术，让我们拭目以待。

New Words and Expressions:

parameter [pə'ræmɪtə]*n.* 参数；系数；参量

optimization [ˌɒptɪmaɪ'zeɪʃən]*n.* 最佳化，最优化

dynamically [daɪ'næmɪkli]*adv.* 动态地；充满活力地；不断变化地

concept ['kɒnsept]*n.* 观念，概念

prototype ['prəutətaɪp]*n.* 原型；标准，模范

workpieces ['wɜrk,pis] *n.* 工件；轧件；工件壁厚

deformation [,diːfɔː'meɪʃ(ə)n]*n.*变形

discharge [dɪs'tʃɑːdʒ] *vt.* 解雇；卸下；放出；免除

conservative [kən'sɜːvətɪv] *adj.*保守的

analytical [ænə'lɪtɪk(ə)l] *adj.* 分析，解析的；善于分析的

prediction [prɪ'dɪkʃ(ə)n] *n.* 预言

constraints [kən'streint] 约束；限制；

algorithms ['ælgərɪ] *n.* [计][数] 算法；算法式

Questions:

Answer the following questions according to the text.

（1）What is the first paragraph of this article?

（2）What are the two cutting parameters?

（3）What is the impact of nature on the algorithm?

Passage Two: Key Technology of High Efficiency Digital Processing for Aircraft Structural Parts

Aircraft structural parts using a large number of integrated design mode. Its structure is complex and the types of features is variety which leading to manufacturing is more difficult. The aircraft structural parts achieve digital manufacturing process after years of technical research and application. But compared with foreign advanced technology, there are still the following problems. The NC programming cycle of large aircraft structural part is long, and the quality of programming is low. In the processing, it is easy to produce deformation. The deformation control is difficult. And the processing efficiency is not high. It is difficult to control the quality of the parts. In short, the processing of aircraft structural parts is still dependent on personnel experience. In the process still needs continuous manual intervention and control which hindering the implementation of automation. Therefore, it is important to develop the research of the automatic and intelligent machining technology in the manufacturing process of aircraft structural parts and realize the flexible manufacture of aircraft structural parts which has great significance to improve the technical level of aircraft manufacturing industry.

1. Integrated process programming

The efficiency and quality of aircraft structural parts process programming has become one of the major bottlenecks in the manufacturing process cycle of aircraft structural parts. Researchers proposed user custom machining feature modeling method oriented complex aircraft structure parts and machining feature recognition technology based on holistic attribute adjacency graph. Meanwhile, they developed an intelligent programming system based on feature which realization the intelligent programming of complex aircraft structure parts. The system can achieve automatic recognition of machining features for aircraft structure parts and aircraft structure piece machining process planning and NC program compiling automatically with process knowledge database. Cutting parameter database and the manufacturing resource database platform, which providing complete solutions for aircraft structural parts programming process. The recognition rate of the typical features is as high as 95%, and the programming efficiency is increased by more than 3 times. With the expansion of the system functions, the system will become an integrated platform, including process examination, feature recognition and process planning, tool path generation, post processing, simulation and optimizing for the whole programming process of aircraft structural parts. Meanwhile, with the continuous improvement and perfection of process knowledge database, cutting parameter database and manufacturing resource database, the future aircraft structural parts programming tasks will be completed all by computer. The Engineer will also be liberated from the arduous process programming tasks, to carry out process technology researching and process knowledge maintenance and updating.

2. Automatic performance evaluation of NC equipment

NC equipment's performance and operation status will direct impact on aircraft structure parts processing quality and processing time. Especially under the request of without manual intervention, the accuracy and performance and failure of the machine tool real-time prediction assessment and analysis and processing put forward higher requirements. The accuracy of the machine tool is an important guarantee for the machining quality of aircraft structural parts. Traditional machine tool accuracy assessment method has the following problems: firstly, the detection time is long. It need to analyze a large number of data. Secondly, it is unable to achieve the real-time online compensation for the precision of the machine tool which need artificial intervention. Germany DST machine tool companies developed a set of machine tool accuracy computer-aided diagnosis system (SQS). The system can automatic detect and compensate the machine tool's accuracy in space and five axis linkage precision fast and calculate and assess of spatial accuracy after the compensation by running a set of predefined standard test procedures. It is mainly through the optical principle to detect the spatial error and R-TEST to detect the attitude error of the rotating head. This machine tool precision automation system developed by DST can complete the automatic detection and compensation of the precision of the machine in 30 minutes. In the intelligent manufacturing plant, it achieve real-time tracking the accuracy of the machine tool. To integrate with the information system of the workshop, we can update the manufacturing resource database in time and provide the selection basis of the machine tool in the manufacturing process of aircraft structure parts. CNC machine tool itself is a very complex system. If it can predict the failure of machine tools and pre-maintain in timely which will greatly improve the utilization rate of machine tools. Integration and fusion in depth of advanced manufacturing technology and information technology and intelligent technology in digital workshop environment achieve intelligent prediction and maintenance of machine tools. Through a sensor network system realizes real-time acquisition of vibration and current and noise and signal of electrical. The sensing signal is transmitted to the control system by industrial Ethernet. According to the received signal, the control system gives the early warning and decision making for the key functional components' health status based on the neural network and other inference algorithms with the support of expert knowledge base. After receiving the decision information, the machine tool adjust its own operation mode adaptively or maintain by maintenance engineer to reduce product quality accidents and lockout caused by the failure of machine tools.

3. Adaptive control technology for machining process

The machining quality and machining efficiency of aircraft structural parts is not only related to machine tools and NC code. But also related to the tools and fixture and measurement and deformation control technology and so on. The adaptive clamping fixture and real-time machining process monitoring and machining and measurement integration technology will be an important part of intelligent flexible production line in the future intelligent machining plants. With the real-time monitoring of the cutting force and the rigid of the part based on in process feature, the adaptive fixture can adjust the fix state and force to match the workpiece and the cutting force to prevent the deformation of the workpiece. At the same time, in order to adapt to the flexible production without artificial intervention mode, the parts was classified family according to the structure characteristics of the aircraft structure. Meanwhile, considering the characteristic of manufacturing process and structure size, the standard quick fixture model is used to realize parts with different specifications and the special fixture machining on different machine tools, in order to adapt to automated production mode of aircraft structural parts in the digital workshop. At present, foreign countries have developed many sets of commercial monitoring software. But machining state recognition only based on monitoring signal. The machining process of complex aircraft structural parts is susceptible to the effects of fluctuations of the machining state which generating frequent false alarms that affect the normal production. Therefore, we need to make in-depth research on the remote monitoring and control of CNC system technology. The spindle cutting power will be real-time monitor to recognize machining state combined with process information. Once found abnormal machining state, it should be control machine tool stop machining immediately to prevent parts quality problem caused by tool wear and breakage and tool path error and so on to enhance recognition accuracy of machining status. Machining and measurement integration technology is mainly used in the parts clamping processing or complete one or some steps machining. The spatial location and feature size of the part were measured with the detection system built-in machine. Then readjust the machine coordinate system and NC code of parts for further machining according to the error distribution by analysis of the measurement data to ensure the dimension requirements of design. The machining quality of the parts in the flexible production line is guaranteed, and the machining error caused by these factors such as the parts loading and clamping and the positioning of tools and parts and the deformation of the parts are reduced.

4. Intelligent production management and control in manufacturing process

Intelligent production management and control center will be the brain of aircraft structural parts machining plant in the future which realization the data collection, analysis, decisionmaking and other functions in one set and responsible for the unified allocation and management of all the resources in the plant. The aircraft structural parts CNC manufacturing process control, job scheduling, on-site management and manufacturing resource management is achieve with the support of digital production management system. On this basis, a virtual factory which real time operation with the actual factory workshop should be built. The production scheduling need be simulated in the virtual plant in order to determine the effectiveness of the scheduling. At the same time, the actual plant operation data is transmitted to the intelligent production control center to drive the virtual factory operation. Intelligent production control center will do decision-making according to the feedback data that the virtual factory simulating to achieve the actual factory scheduling control. The intelligent production management and control center effectively extends the function of the human brain, and has a great effect on the realization the without manual intervention or less manual intervention manufacturing of the aircraft structural parts.

参考译文：飞机结构件高效数字处理关键技术

飞机结构件采用大量综合设计模式，其结构复杂，特征种类多样，导致了制造困难。虽然飞机结构件经过多年的技术研究和应用，实了现数字化制造工艺，但与国外先进技术相比，仍存在以下问题。大型飞机结构部件的NC编程周期长，编程质量低；在加工过程中，容易产生变形，变形控制困难，处理效率不高，很难控制零件的质量。总之，飞机结构件的加工依然依赖于人员的经验。在此过程中仍然需要持续的手动干预和控制，这阻碍了自动化的实施。因此，开展飞机结构件制造过程中自动智能加工技术的研究，实现飞机结构件的灵活制造，对提高飞机制造业的技术水平具有重要意义。

1. 整合整个飞机结构件编程过程

编程的效率和质量已成为飞机结构件制造工艺周期的主要瓶颈之一。研究者提出了以用户定制加工特征的建模方法指导复杂飞机结构件和基于整体属性相邻关系的零件加工特征识别技术。同时，开发了基于某些功能的智能编程系统，实现了复杂飞机结构部件的智能编程。该系统可以自动识别飞机结构零件,完成飞机结构件加工过程规划，实现基于过程知识数据库的数控程序自动加工。切削参数数据库和制造资源数据库平台能为飞机结构零件编程过程提供完整的解决方案。典型特征识别率高达95％，编程效率提高3倍以上。随着系统功能的扩展，系统将检测飞机结构件整个编程过程，如特征识别和过程规划，刀具路径生成，后处理，仿真和优化等集成为一个平台。同时，随着切削参数数据库和制造资源数据库等知识数据库的不断完善，未来飞机结构件编程任务将全部由计算机完成。工程师也将从繁琐的编程任务中解脱出来，进行技术研究和知识的维护与更新。

2. 数控设备的自动性能评估

数控机床的性能和运行状态将直接影响飞机结构件的加工质量和加工时间。特别是在无需人工干预的要求下，对机床持续实时预测评估和分析处理的准确性及性能提出了更高的要求。机床的精度是加工过程中对飞机结构件质量的重要保证。传统机床精度评估方法存在以下问题：一是检测时间长，它需要分析大量的数据。其次，无法满足人工干预机床精度的实时在线补偿的需要。德国DST机床公司开发了一套机床精度计算机辅助诊断系统（SQS）。该系统可以自动检测和补偿机床的空间精度和五轴联动精度，通过运行一套预定义的标准测试程序，对补偿后的空间精度进行计算和评估。该系统主要是通过光学原理检测空间误差和利用R-TEST来检测旋转头的姿态误差。DST开发的这种机床精密自动化系统可以在30分钟内完成机器精度的自动检测和补偿。在智能制造工厂，该系统实现了机床的实时跟踪。为了与车间的信息系统整合，我们可以及时更新制造资源数据库，并为飞机结构部件制造过程中的机床选型提供依据。数控机床本身是一个非常复杂的系统，如果能够预测机床的故障，及时预先维护，将大大提高机床的利用率。数字车间环境中先进制造技术、信息技术、智能技术的深度融合，实现了机床的智能预测和维护。通过传感器网络系统实现电气振动和电流、噪声和电信号的实时采集。感测信号通过工业以太网传输到控制系统，根据接收到的信号，控制系统在专家知识库的支持下，基于神经网络和其他推理算法，对关键功能组件的健康状况进行预警和决策。收到确认信息后，机床自主调整自己的运行模式或由维护工程师维护，以减少机床故障引起的产品质量事故。

3. 加工过程的自适应控制技术

飞机结构件的加工质量和加工效率不仅与机床和数控代码相关，而且还涉及工具和夹具以及测量和变形控制技术等。自适应夹具和实时加工监控过程以及加工和测量集成技术将成为未来智能加工厂智能柔性生产线的重要组成部分。实时监测切削力和零件刚性，自适应夹具可以调整固定状态和装夹力以匹配工件和切削力，从而防止工件变形。为了适应柔性的生产，而不需要人工干预其工作，根据飞机的结构特点，将部件归类，同时，考虑到制造工艺和结构尺寸的特点，标准快速夹具模型将用于实现不同规格的零件和不同机床上的专用夹具加工，以适应数字化飞机结构件的自动化生产模式。目前，国外已开发出多套商业监控软件，但加工状态识别只能基于监控信号。复杂飞机结构件的加工过程易受加工状态波动的影响，会频繁产生影响正常生产的虚假报警。因此，我们需要对数控系统技术的远程监控进行深入的研究。主轴切割功率将被实时监控，以识别与加工信息相结合的加工状态。一旦发现异常加工状态，应立即控制机床停止加工，防止刀具磨损和刀具路径误差等引起的零件质量问题，从而提高加工状态的识别精度。加工和测量集成技术主要用于零件夹紧加工或完成一个或多个步骤加工。使用检测系统内置机器测量零件的空间位置和特征尺寸。然后通过分析测量数据，根据误差分布重新调整机床坐标系和零件代码，进一步加工，以确保设计的尺寸要求。柔性生产线中零件的加工质量得到保证，并减少了部件加载和夹紧以及刀具和零件定位以及部件变形等因素造成的加工误差。

4. 智能生产管理与制造过程

智能生产管理和控制中心将成为未来飞机结构件加工厂的大脑，集数据采集、分析、决策等功能于一体，负责统一配置和管理工厂的所有资源。飞机结构件数控制造过程控制、作业调度、现场管理和制造资源管理都是在数字化生产管理系统的支持下实现的。在此基础上，建立与实际工厂车间实时运行的虚拟工厂。在虚拟工厂中需要对生产调度进行模拟，以确定调度的有效性。同时，将实际的工厂运行数据传输到智能生产控制中心，推动虚拟工厂运行。智能生产控制中心将根据虚拟工厂模拟的反馈数据做出决策，实现实际的工厂调度控制。智能生产管理和控制中心有效扩展了人脑的功能，对实现无需人工干预或者需要较少手动干预的飞机结构部件的制造起到很大的作用。

New Words and Expressions:

digital ['dɪdʒɪt(ə)l]*n.* 数字；键

structural ['strʌktʃ(ə)r(ə)l] *adj.* 结构的；建筑的

NC (numerical control) 数控机床

programming ['prəʊgræmɪŋ] *n.*设计，规划；编制程序，[计] 程序编制

deformation [ˌdiːfɔːˈmeɪʃ(ə)n] *n.* 变形

implementation [ɪmplɪmenˈteɪʃ(ə)n] *n.* [计] 实现；履行；安装启用

bottlenecks ['bɒtlnek] *n.* [包装] 瓶颈

oriented ['ɔːrɪentɪd] *adj.* 导向的；定向的；以……为方向的

adjacency [əˈdʒeɪsənsɪ] *n.* 毗邻；四周；邻接物

database ['deɪtəbeɪs] *n.* 数据库，资料库

examination [ɪɡˌzæmɪˈneɪʃ(ə)n; eg-] *n.* 考试；检查；查问

simulation [ˌsɪmjʊˈleɪʃən] *n.* 仿真；模拟；模仿；假装

axis ['æksɪs] *n.* 轴；轴线；轴心国

linkage ['lɪŋkɪdʒ] *n.* 连接；结合；联接；联动装置

R-TEST 或倒转检验, 统计学, r-检验

adaptive [əˈdæptɪv] *adj.* 适应的，适合的

workpiece ['wɜːk,piːs] *n.* 工件；轧件；工件壁厚

specifications [spɛsəfəˈkeʃənz] *n.*规格；说明书

fluctuations [ˌflʌktjuˈeiʃəns] *n.* [物] 波动；变动；起伏现象

decisionmaking [diˈsiʒənˌmeikiŋ] *n.* 决策

Questions:

Answer the following questions according to the text.

（1）What are the major bottlenecks in aircraft programming?

（2）What are people called automatic performance evaluation and safety technology CNC machine tools ?

（3）Is the adaptive control technology of the process related to that of several technologies?

Unit Eight

Passage One: The Drill in Aircraft Assembly

Aircraft assembly operations consume up to 40% of the overall manufacturing cost of aircraft. The reason why such a high proportion is because the number of holes behind this requires a lot of labor costs.

The current method of hole drilling and countersinking for fastener installation uses unique drill fixtures (DFs) for each assembly and subassembly to be processed, at an average cost of $50,000 each. Well over 900 conventional drill fixtures are required for airframe assembly on each program. All aircraft production programs have similar requirements for hand drilling operations.

The application of the drill tool to the assembly tool is temporary, and installation and removal are required at each assembly. The installation and removal of the drill tool exposes the tool and the part to poten-tial damage and, in many cases, requires the use of handling equipment. The assembly tool is designed to accommodate the attachment of the drill tools to them by the use of pins and bolts.

When the drill tools are applied to the assembly fixture, they are attached in such a way as to hold them approximately 1/4 away from the surface of the part. This is done in order to allow the chip to clear as the hole is drilled in the part. When the part is composed of a metallic substructure and a composites skin, the chip is trapped between the drill tool and the part. The metallic part of the chip spins against the composite skin and can cause damage to the composite face of the skin. Since the variance in the parts causes variances in their placement within the assembly fixture. This effects the position of the hole to the substructure and the angle of the hole to the contour of the part.

The countersink pilot and the subsequent countersunk hole derive their relationship to the contour of the part from the drilled hole. Therefore, if the hole is not normal to contour, the countersunk hole will be out of round and either too deep or too shallow proportionally. In addition, once the relationship of the hole and the countersink are established to the contour, the fastener is commit fed to follow the path of the hole and the countersink. In short, if the hole is out of contour, then the rest of the process follows to the degree of error established by the hole.

This complex process is meant to control the position of the holes, countersinks and fastener. Further complicating the process are the standard tools which are hand held, air powered and manually fed. Speeds are controlled by the air line pressure and the operator. Feeds are controlled by the size, weight and strength of the individual pushing on the hand-held drill motor.

参考译文： 飞机装配中的钻孔

飞机装配业务占飞机总体制造成本的40％。之所以占这么高的比例，是因为这大量的钻孔数目需要大量的劳动成本。

目前的紧固件安装孔钻孔和锪孔方法是使用专用钻孔夹具（DF）来处理每个装配组件，平均每个成本为50 000美元。每个程序上的机身装配都需要900多个常规钻具。所有的飞机生产流程，对手工钻的操作都有类似要求。

将钻头应用于装配工具是临时的，每个装配都需要安装和拆卸。钻头的安装和拆卸使工具和工件存在潜在的损坏危险，并且在许多情况下需要使用装卸设备。装配工具被设计成销和螺栓来完成钻头对其的附接。

当钻具被施加到装配夹具上时，它们要保持离开部件的表面约1/4的距离的方式被附接，这是为了允许钻屑清除，因为钻孔是由部分金属底座和复合材料表皮组成，钻屑容易被夹在钻具和零件之间，钻屑的金属部分与复合材料表层相互旋转，可能会损坏复合材料的表面。由于工件的变化导致装配夹具中的位置变化，这影响到了孔到底座的位置和孔与部件轮廓的角度。

埋头孔导杆和随后的沉头孔与钻孔的部件轮廓相关，因此，如果孔不符合轮廓，则埋头孔将不成圆形，要么成比例地陷入太深，要么太浅。另外，一旦将孔和埋头孔相对于轮廓的关系确定，则紧固件就会沿着孔和埋头孔的路径进给。简而言之，如果孔不在轮廓上，则后续的工序就会沿着孔的误差而继续。

这个复杂的过程是为了控制孔、埋头孔和紧固件的位置。进一步复杂的过程是手动钻头、空气动力和手持进给。进给速度是通过空气管路压力和操作员来控制的。进给量则通过对手持式钻头电机不同的大小、刚度和强度单独控制达到。

New Words and Expressions:

mechanization [ˌmekənaɪˈzeɪʃən] n.机械化；机动化

automation [ɔːtəˈmeɪʃ(ə)n] n.自动化；自动操作

disproportionate [ˌdɪsprəˈpɔːʃ(ə)nət] adj / adv. 不成比例的；不成比例地

current [ˈkʌr(ə)nt] adj. 现在的；流通的，通用的；最近的

iterations [ˌitəˈreiʃən] n. 迭代次数；反复（iteration的复数）

fiberglass [ˈfaɪbərˌɡlæs] n. 玻璃纤维；玻璃丝

accommodate [əˈkɒmədeɪt] vt. 容纳；使适应；供应；调解

variations [ˌveəriːˈeiʃənz] n. 变奏曲，变更；（variation的复数形式）

countersink [ˈkaʊntəsɪŋk] n. 埋头孔；暗钉眼；皿锥

error [ˈerə] n. 误差；错误；过失

Questions:

Answer the following questions according to the text.

（1）What is the hole drilling and countersinking method？

（2）What is the composition of the hole？

（3）What are the further controls on the operation of the hole？

Passage Two: The Positioning System for Aircraft Structural Assembly

Automated assembling of large aircraft components offers an alternative solution for aircraft manufacturers to cut costs, improve process quality, and shorten time to market. More and more, flexible and automated systems are replacing hard system throughout the process of aircraft structural assembly, for instance, in the manufacturing of the fuselage sections from shell panels, the alignment of the sections to build the fuselage, and the joining of wings and tail units to the fuselage. The diagrammatic sketch of aircraft structual assemldy can be sean in Fig.8-1.

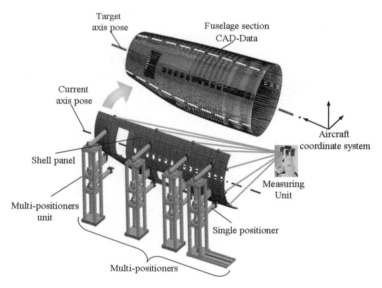

Fig .8-1   Aircraft structural assembly

In the past decade, a few approaches were proposed for automated assembly systems consisting of multiple manipulators, which act simultaneously to handle the aircraft components. Especially in conjunction with the use of innovative composite materials ,the challenge for these assembly systems consists in the proper handling and accurate positioning of the aircraft components.

The Eco Positioner from Dürr Systems GmbH is a new conception of a modular and reconfigurable positioning technology that provides a holistic solution, taking into consideration the positioning task as well as the operational conditions and environmental influences. It is a mechanical system consisting of multi-manipulators, or multi-positioners, that pick up the aircraft component at defined attachment points and manipulate it in six degrees of freedom (three translational and three rotational).

The Eco Positioner includes a measurement module, a large metrology system and measurement software that provides information on the current and ideal poses (positions and orientations) of the aircraft component. The current and the ideal poses are determined according to the reference frame related to the CAD models of the components and the aircraft structure.

A reliable robot-control platform guarantees a high synchronization of up to 48 actuated axes per control unit. The EcoPRC control system is well-proven worldwide and is used in more than 6,000 robot applications. With the aid of force sensors mounted on the end-effector of each positioner, the forces acting on the aircraft component can be monitored during the motion. This ensures strain-free handling of the components and thus prevents them from uncontrolled deflection.

The positioning of components and working tools is the essential task in the assembly of fuselage sections from single shell panels. For large and geometrically complex structural components, which by their very nature suffer from considerable physical distortion, the positioning task is even more crucial.

Classical assembly methods use large, complex jigs, which are specially tailored for an aircraft type. The jigs physically control the shape of the aircraft component and have to be moved together with the component to the assembly position. An advantage of hard jigs is that they do not require highly skilled workers; however, the jigs are expensive, have long manufacturing lead times, and cannot be modified economically for use on other aircraft types.

An alternative to hard jigs is the use of flexible manipulators. Combined with an accurate, large metrology system and an advanced control system, the positioning task can be structured in the following sequence:

（1）Pick up components using multi positioners at defined attachment points located on the component.

（2）Measure the geometry of the component using, for instance, a laser tracker and spherical-mounted retro-reflectors that are placed at defined locations.

（3）Adjust the component geometry with respect to product design (CAD data) using, for instance, Best-Fit strategies. This operation is sometimes necessary, especially for CFRP components.

（4）Measure the current pose of the component.

（5）Generate the motion path from current to ideal pose as corresponding to the CAD data of the product design. Advanced controls with robust path planning strategies are mandatory for this operation to avoid damage to the component.

（6）Move the component to ideal pose, taking into account the restrictions on the accuracy and the synchronicity of motion.

Four functionalities are needed to carry out the assembly task. First, an accurate measurement of points specifically placed on the component to recognize its shape and determine its pose must be taken. Various metrology systems can be used—for instance, systems based on laser or combined laser and IR technology such as an indoor global positioning system.

Second, acquisition, analysis, and processing of the measurement data must be ensured. This can be done via metrology software, which can analyze the measurement data and compare them with the CAD data of the components and their locations in the assembly site using robust Best-Fit strategies.

The third task consists of the generation of the trajectory along which the component will be moved from current to ideal pose. Since multiple positioners simultaneously act on the component, this task constitutes a tremendous challenge for any control system. For path generation, the control system must not only account for the restrictions related to the workspace of each positioner but also avoid any excessive strain induced by the positioner on the component during the motion.

The fourth and most crucial task is execution of the motion. Despite all additional impact in terms of external loads and environmental conditions such as temperature fluctuation, the single positioner should feature a positioning accuracy superior to that required for the component itself.

All in all, this assembly system requires sophisticated and collaborative collaboration to complete.

参考译文：飞机结构装配定位系统

大型飞机部件的自动装配是为了降低飞机制造成本，为提高加工质量和缩短上市时间提供了一种解决方法。越来越多的柔性和自动化系统正在替换飞机结构组装过程中的硬性系统，例如，在从壳板制造机身部分时，通过部件的对准来构建机身，将机翼和尾翼单元连接到机身。图8-1为飞机结构装配示意图。

在过去十年中，提出了一种自动组装系统的方法，该系统由多个操纵器组成，这些操纵器同时处理飞行器部件。特别是使用创新的复合材料时，这些组装系统的挑战在于妥善处理并准确定位飞机部件。

来自杜尔系统的Eco定位器用的是一种模块化和可重构的定位技术的新概念，提供了一个整体的解决方案，同时考虑到定位任务以及操作条件和环境影响。它是一个机械系统，由多个操纵器或多位定位器组成，可以将其定义在连接点上，并以六个自由度（三个平移和三个旋转）进行操纵。

Eco定位器包括一个测量模块，一个大型计量系统和测量软件，提供有关飞机部件当前和理想姿态（位置和方向）的信息。目前，理想姿态是根据与部件和飞机结构相关的CAD模型的参考系确定。

一个可靠的机器人控制平台可保证多达每个控制单元48个驱动轴的高同步性。EcoPRC控制系统在世界范围内得到了广泛的验证，并被应用于6 000多个机器人中。借助于安装在每个定位器末端执行器上的力传感器，可以在运动过程中监视作用在飞行器部件上的力。这确保了组件的无应变处理，从而防止它们不受控制的偏转。

工件和刀具的定位是单壳板组装机身部分的基本任务。对于大型和几何复杂的结构部件，由于其本身具有相当大的物理扭曲，定位任务更为关键。

传统的组装方法是使用大型复杂的专门针对飞机的夹具。夹具控制飞机部件的形状，并且必须与工件一起移动到组装位置。硬性夹具的优点是不需要高技能工人，然而，夹具昂贵，制造时间长，并且不能通过简易地修改来应于其他飞机类型。

硬性夹具的替代方案是使用柔性操纵器。结合精准的大型计量系统和先进的控制系统，定位任务可以按以下顺序进行构造：

（1）使用位于工件上定义的附件点，使用多位定位器夹持工件。

（2）使用放置在规定位置的激光跟踪器和球形反射器来测量工件的几何形状。

（3）使用最优位置策略，结合相关产品设计数据（CAD数据）调整工件几何尺寸。这种操作有时是必要的，特别是对于碳纤维复合工件。

（4）测量工件的当前位姿。

（5）根据产品设计的CAD数据生成从当前位姿到理想位姿的运动路径。为了避免损坏工件，必须强制采取具有强大路径规划策略的高级控件。

（6）将组件移动到理想姿态，同时考虑到运动精度和位移同步性的限制。

需要四个功能来执行组装任务。

第一，必须准确测定工件指定点，以确定工件的形状和位置。使用各种计量系统，例如基于激光、组合激光、红外技术的全球定位系统。

第二，要确保测量数据的采集、分析和处理。这可以通过计量软件完成，它可以分析测量数据，并使用强大的最佳拟合策略将其与组件的CAD数据在装配现场的位置进行比较。

第三项任务包括生成组件，并生成从当前位置移动到理想位置的轨迹。由于多个定位器同时作用于组件，所以这个任务对任何控制系统都构成了巨大的挑战。对于路径生成，控制系统不仅要考虑与每个定位器的工作空间限制，而且还要避免在运动期间，定位器在部件上引起的任何过度的应变。

第四个也是最重要的任务是执行方案。尽管在外部负载和环境条件（如温度波动）方面有外在的影响，单一定位器应具有优于组件本身所需的定位精度。

总而言之，这种装配系统需要精准且多个方面的协同合作才可以完成。

New Words and Expressions:

automated ['ɔ:tə‚meɪtɪd] *adj.* 自动化的；机械化的

alternative [ɔ:l'tɜ:nətɪv; ɒl-] *adj.* 供选择的；选择性的；交替的

fuselage ['fju:zəlɑ:ʒ; -lɪdʒ] *n.* [航] 机身（飞机）

accurate ['ækjərət] *adj.* 精确的

reconfigurable [ri'kɒnfigərəbl] *adj.* 可重构的

mechanical [mɪ'kænɪk(ə)l] *adj.* 机械的；力学的；手工操作的

translationa [trænz'leʃənl] *adj.* 平移的，直移的

rotational [rou'teʃənl] *adj.* 转动的；回转的；轮流的

synchronization [‚sɪŋkrənaɪ'zeɪʃən] *n.* [物] 同步；同时性

conception [kən'sepʃ(ə)n] *n.* 怀孕；概念；设想；开始

distortion [dɪ'stɔːʃ(ə)n] *n.* 变形；[物] 失真；扭曲；曲解

tailored ['teɪləd] *adj.* 定做的；裁缝做的；剪裁讲究的

CFRP *abbr.* 碳纤维复合材料（Carbon Fiber Reinforced Plastics）

trajectory [trə'dʒekt(ə)rɪ; 'trædʒɪkt(ə)rɪ] *n.* [物] 轨道，轨线；弹道

Questions:

Answer the following questions according to the text.

（1）What is the innovation of the method of automatically assembling the system?

（2）How many brake weeks can the robot control platform control?

（3）What are the external factors that affect the fourth function?

Unit Nine

Passage One: An Introduction to Unmanned Aerial Vehicles

The term UAV is an abbreviation of Unmanned Aerial vehicle, meaning aerial vehicles which operate without a human pilot. UAVs are commonly used in both the military and police forces in situations where the risk of sending a human piloted aircraft is unacceptable, or the situation makes using a manned aircraft impractical.

One of the predecessors of today's fully autonomous UAVs were the "aerial torpedoes", designed and built during World War One. These were primitive UAVs, relying on mechanical gyroscopes to maintain straight and level flight, and flying until they ran out of fuel. They would then fall from the sky and deliver and explosive payload.

More advanced UAVs used radio technology for guidance, allowing them to fly missions and return. They were constantly controlled by a human pilot, and were not capable of flying themselves. This made them much like today's RC model airplanes which many people fly as a hobby.

After the invention of the integrated circuit, engineers were able to build sophisticated UAVs, using electronic autopilots. It was at this stage of development that UAVs became widely used in military applications. UAVs could be deployed, fly themselves to a target location, and either attack the location with weapons, or survey it with cameras and other sensor equipment.

Modern UAVs are controlled with both autopilots, and human controllers in ground stations. This allows them to fly long, uneventfully flights under their own control, and fly under the command of a human pilot during complicated phases of the mission.

Since their creation, UAVs have found many uses in police, military, and in some cases, civil applications. Currently, UAVs are most often used for the following tasks:

（1）Aerial Reconnaissance—UAVs are often used to get aerial video of a remote location, especially where there would be unacceptable risk to the pilot of a manned aircraft. UAVs can be equipped with high resolution still, video, and even infrared cameras. The information obtained by the UAV can be streamed back to the control center in real time.

（2）Scientific Research—In many cases, scientific research necessitates obtaining data from hazardous, or remote locations. A good example is hurricane research, which often involves sending a large manned aircraft into the center of the storm to obtain meteorological data. A UAV can be used to obtain this data, with no risk to a human pilot.

（3）Logistics and Transportation—UAVs can be used to carry and deliver a variety of payloads. Helicopter type UAVs are well suited to this purpose, because payloads can be suspended from the bottom of the airframe, with little aerodynamic penalty.

There are many different types of UAVs, designed for different purposes. The US air force is one of the most prominent users of UAV technology, and classifies UAVs by dividing them into tiers.

Some UAVs use a block design, and are well suited to carrying large amounts of cargo. The first UAVs were called "drones" and were not autonomous, because they required constant control input from a remote human pilot. Computer technology now allows UAVs to make their own decisions, or fly autonomously. Autonomous flight involves the UAV making decisions as it flies.

Generally, autonomous flight consists of the following operations:

（1）Interpreting sensor input, and merging the input of multiple sensors.

（2）Communicating with ground stations, satellites, and other UAVs and aircraft.

（3）Determining the ideal course to fly for a given mission, based on sensor input.

（4）Determining the best maneuvers to perform for a given task.

（5）In some cases, cooperating with other UAVs to accomplish a common task.

UAVs represent an area of rapid development in both military and civilian applications.

UAVs unique capability of flying dangerous, long, or precision missions give it a unique advantage over conventional aircraft.

参考译文：无人机简介

UAV是无人机的缩写，是指无人驾驶运行的空中飞行器。无人机通常用在驾驶员存在不能接受的风险情况或者使用有人驾驶飞机不切实际的情况下，在军事和警察部队中常常使用。

现今完全自动的无人机的前辈之一是在第一次世界大战期间设计和建造的"空中鱼雷"。这些是初始的无人机，依靠机械陀螺仪来保持直线和水平飞行，直到它们用完了燃油。然后他们将从天空中掉下来，发出有毒有害物质。

更先进的无人机使用无线电技术进行指导，允许它们完成飞行任务并返回。它们不断受到人类操控员的控制，不能自主飞行。这使得它们更像今天飞行爱好者的RC模型飞机。

在集成电路发明之后，工程师们能够使用电子自动驾驶仪来构建复杂的无人机。在这个发展阶段，无人机在军事应用中广泛应用。无人机可以部署飞行到目标位置，并用武器攻击该地点，或用相机和其他传感器设备进行测量。

现代无人机由地面站中的自动驾驶仪和人类控制员控制。这允许它们在自己的控制下平稳地长时间飞行，并在人类操控员的指挥下进行复杂任务。

无人机自创建以来，在公安、军事以及某些情况下都发现了很多用途。目前，无人机最常用于以下任务：

（1）空中侦察——无人机通常用于获取远程位置的空中视频，特别是在有人驾驶将面临不可接受的风险的状况下。无人机可以配备高分辨率静态图片、视频甚至红外摄像机。无人机所获得的信息可以实时回传到控制中心。

（2）科学研究——在许多情况下，科学研究需要从危险或偏远地区获取数据。一个很好的例子就是飓风研究，这通常涉及将大型载人飞机送入风暴中心，以获得气象资料。无人机可用于获取此数据，而不会对人类飞行员造成任何风险。

（3）物流与运输——无人机可用于运载和传输各种有效载荷。直升机型无人机非常适合于此应用，因为有效载荷可以从机身的底部悬挂，而几乎没有任何空气动力学上的障碍。

有许多不同类型的无人机，被设计用于不同目的。美国空军是无人机技术最突出的用户之一，它将无人机分为几个级别。

一些无人机使用分块设计，非常适合携带大量货物。第一个无人机被称为"雄峰"，因为他们需要来自远程人类操控员的持续控制输入，所以并不是全自动的。计算机技术现在允许无人机做出自己的决定，自主飞行。自动无人机涉及到无人机飞行时需要做出决定。

一般来说，自主飞行包括以下操作：

（1）解释并合并多个传感器的输入。

（2）与地面站、卫星和其他无人机和飞机通信。

（3）根据传感器输入确定飞行任务的理想航向。

（4）为给定执行任务确定最佳操作。

（5）在某些情况下，与其他无人机合作完成一项共同任务。

无人机是军事和民用领域迅速发展的领域。无人机独有的飞行危险、长时时效性和完成任务的精确性，使之比传统飞机具有了更加独特的优势。

New Words and Expressions:

abbreviation [əbriːvɪˈeɪʃ(ə)n]n. 缩写；缩写词

situations [ˌsɪtʃʊˈeɪʃnz]n. 状况；情境；局面

unacceptable [ʌnəkˈseptəb(ə)l]adj. 不能接受的；不受欢迎的

predecessors [ˈpriːdɪˌsɛsə]n. 前任

gyroscopes [ˈdʒaɪrəˌskəup]n. 陀螺仪

sophisticated [səˈfɪstɪkeɪtɪd]adj. 复杂的；精致的；久经世故的；富有经验的

weapons [ˈwɛpən]n. 武器；武装切换；斗争工具

Reconnaissance [rɪˈkɒnɪs(ə)ns]n. [军] 侦察；勘测；搜索；事先考查

necessitate [nɪˈsesɪteɪt]vt. 使成为必需，需要；迫使

hurricane [ˈhʌrɪkən]n. 飓风，暴风

meteorological [ˌmiːtɪərəˈlɒdʒɪkəl]adj. 气象的；气象学的

aerodynamic [ˌɛərodaɪˈnæmɪk]adj. 空气动力学的，[航] 航空动力学的

drones [drəʊn]n. 雄蜂；（英）懒汉；遥控飞机

capability [ˌkeɪpəˈbɪləti]n. 才能，能力；性能，容量

Questions:

Answer the following questions according to the text.

(1) According to the first part explain what is the UAV?

(2) What are the tasks for UAVs?

(3) What is the UAV's own flight process?

Passage Two: The Autopilot of Unmanned Aerial Vehicle

Unmanned aerial vehicles (UAVs) are becoming more and more popular in a wide field of applications nowadays. Although being developed mainly for military purposes. In the past, it becomes that there are a lot of other areas where they might prove useful. Consider agriculture, where they may be used for field observations or for chemicals distribution. They can patrol over wide forest areas as a fireguard, or they can be used for traffic observation in the cities. In cartography, small UAVs might be used for automatic landscape photographing, being much more cost efficient compared with traditional aerial snap shooting. UAVs are also very interesting for the academic research, as they can be used for various purposes—as flying laboratories, proving ground for control algorithms, or as an education tools for students.

Therefore, there is a growing demand for UAV control systems, and many projects—either commercial or academic destined to design an UAV autopilot were (and are) held recently. A lot of impressive results had already been achieved, and many UAVs, more or less autonomous, are used by various organizations. Sadly, they are still unattainable for many potential users, as products are very expensive, and academic researchers are often reluctant to give away specific construction details of their inventions. Many teams had published their results, but none of them had released full technical documentation of their control systems for public usage. Hence, every newcomer to this field is compelled to start his project from scratch, again and again. The complexity of the whole design process may discourage many of them, and a lot of time and resources are wasted by reinventing things already known to more experienced teams. That is why we designed RAMA.

RAMA stands for Remotely operated Aerial Model Autopilot. It is an open project, held at the Department of Control Engineering, Faculty of Electrical Engineering of the Czech Technical University in Prague, which purpose is to design a universal lightweight and compact control system for small UAVs. RAMA is intended to be used mainly in the academic environment, and shall serve as a good starting point for anyone willing to build his own UAV. Main distinctive feature of the RAMA project is that it is totally open, meaning that the whole technical documentation—including wiring diagrams, PCBs (Printed Circuit Boards), software source codes, controller designs, mathematical models and real flight data from our UAV. Therefore, anybody with necessary technical background should be able to build his own RAMA system for whatever purpose, saving a lot of work he would otherwise spend developing his own system.

From engineering point of view, UAV autopilot design poses a lot of challenges. It is very complex multidisciplinary process, covering disciplines from hardware design, sensors and measurement, programming, networking, etc. to mathematic modeling and control theory, artificial intelligence, image and signal processing etc. Therefore it is very interesting for researchers from various fields, and there is still plenty of room for improvements and new approaches, as this field is relatively fresh and unexplored.

The RAMA project had been running since 2004 and had already brought some interesting results. From hardware design and system programming point of view, it is almost complete by now. Currently, extensive work is being carried out on the mathematical modeling and controller designs. A couple of stabilizing controllers had already been introduced and tested with promising results, and dozens of test flights were performed. The project is comprehensively documented at the project website.

A lot of work had already been done in the UAV field by other teams. Most interesting works in the academic environment were proposed by following teams: team from ETH Zurich presented a stabilizing controller for a rotorcraft in . At the University of California in Berkeley a military UAV rotorcraft was developed in the course of the DARPA project . At the MIT (Massachusetts Institute of Technology), they developed a complex non-linear mini-helicopter mathematic model and verified it successfully using two small rotorcrafts.

From non-academic projects, let us at least mention the Yamaha RMAX R-UAV , probably the most advanced commercial UAV currently available at the market. It is a completely autonomous R-UAV, used widely all around the world in such projects as environmental observations, atomic plant surveillance, infrastructure maintenance etc.

参考译文：无人机的自动驾驶仪

如今，无人机（UAV）虽然主要用于军事目的，但是在各个领域也得到了广泛应用。在过去的时间里无人机被证明在很多领域是实用的。在农业方面，它们可用于实地观察或化学药品分布。它们可以在广阔的森林地区巡逻防火，也可以用于城市交通观测。在制图领域，小型无人机可能被用于自动景观拍摄，与传统的空中拍摄相比，该方法成本更低。无人机也可以用作学术研究，因为它们可以用于各种实验目的，如飞行实验室，控制算法的证明，或作为学生的教学工具。

因此，人们对无人机控制系统的需求不断增长，许多项目——无论是商业还是学术界，都是为了设计无人机自动驾驶仪而进行的。许多令人印象深刻的成果已经实现，许多无人机或多或少地通过自动控制被各种组织使用。但是，因为产品非常昂贵，对于许多潜在用户来说，它们难以满足要求，但学术研究人员往往不愿意公布他们发明的详细细节。许多团队已经公布了其结果，但是他们并没有对公众发布完整的控制系统技术文档供大众使用。因此，这个领域的每一个新手都被迫一次又一次地从零开始他的项目。整个设计过程的复杂性可能会阻碍他们，浪费了大量的时间和资源重新创造对于更有经验的团队已知的事情。这就是为什么我们设计RAMA。

RAMA也就是遥控飞行模型自动驾驶仪。它是一个开放的项目，在布拉格的捷克技术大学电气工程学院控制工程系提出，目的是为小型无人机设计一个通用轻巧紧凑的控制系统。 RAMA主要用于学术领域，作为任何一个愿意搭建自己的无人机的人的起点，RAMA项目的主要特点是完全开放，这意味着整个技术文档——包括接线图，PCB（印刷电路板），软件源代码，控制器设计，数学模型和来自我们无人机的实际飞行数据。因此，任何有一定技术背景的人都应该能够根据自己的目的建立自己的RAMA系统，在开发他自己的系统上节省大量的工作。

从工程的角度来看，无人机自动驾驶仪设计面临着许多的挑战。它是非常复杂的多学科过程，涵盖从硬件设计、传感器和测量、编程、网络到数学建模与控制理论、人工智能、图像和信号处理等方面的学科。因此，来自各个领域的研究人员有很大的改进及创新空间，因为这个领域是比较新的且未开发的。

RAMA项目自2004年以来一直在运行，并已经带来了一些有趣的结果。从硬件设计和系统编程的角度来看，现在几乎已经完成了。目前，工作主要针对数学建模和控制器设计，已经引入和测试了几个稳定控制器，获得了有希望的结果，并进行了数十次测试飞行。项目网站上全面地记录了该项目。

其他团队在无人机领域已经做了很多工作。在学术环境中最有趣的作品是由以下团队提出的：苏黎世大学团队提出了一种旋翼航空器的稳定控制器。在DARPA项目的过程中，加利福尼亚大学伯克利分校开发了一种军用无人机旋翼航空器。麻省理工学院开发了一种复杂的非线性微型直升机数学模型，并使用两个小型旋翼飞机成功验证了它们。

从非学术项目，我们可以提到雅马哈RMAX R-UAV，它可能是目前市面上最先进的商用无人机。它是一个完全自主的R-UAV，在环境观测、原子能监测、基础设施维护等项目中广泛应用在世界各地。

New Words and Expressions:

Autopilot ['ɔ:təʊ,paɪlət]*n*. [航] 自动驾驶仪

Unmanned [ʌn'mænd]*adj*. 无人的；无人操纵的；被阉割的

Networked ['netwə:kt]*adj*. 网路的；广播电视联播的

hierarchical [haɪə'rɑːkɪk(ə)l]*adj*. 分层的；等级体系的

agriculture ['ægrɪkʌltʃə] *n*. 农业；农耕；农业生产；农艺，农学

cartography [kɑ'tɒgrəfɪ] *n*. 地图制作，制图；制图学，绘图法

snapshoot ['snæpʃuːt] *vt*. 拍……之快照；快镜拍摄

laboratory ['læbrə,tori] *n*. 实验室

documentation [,dɒkjʊmen'teɪʃ(ə)n] *n*.文件, 证明文件, 史实, 文件编制

extensive [ɪk'stensɪv; ek-] *adj*.广泛的；大量的；广阔的

stabilizing ['stebl,aɪzɪŋ]*v*. 使安定（stabilize的ing形式）

Questions:

Answer the following questions according to the text.

（1）What aspects of UAV can be applied?

（2）What does RAMA mean?

（3）Please describe the research results of several teams.

Unit Ten

Passage One: Academic conference

An academic conference or symposium is a conference for researchers (not necessarily academics) to present and discuss their work. Together with academic or scientific journals, conferences provide an important channel for exchange of information between researchers.

Conferences are usually composed of various presentations. They tend to be short and concise, with a time span of about 10 to 30 minutes, presentations are usually followed by a discussion. The work may be bundled in written form as academic papers and published as the conference proceedings.

Usually a conference will include keynote speakers (often, scholars of some standing, but sometimes individuals from outside academia). The keynote lecture is often longer, lasting sometimes up to an hour and a half, particularly if there are several keynote speakers on a panel.

In addition to presentations, conferences also feature panel discussions, round tables on various issues, poster sessions and workshops. Some conferences take more interactive formats, such as the participant driven "unconference" or various conversational formats.

Prospective presenters are usually asked to submit a short abstract of their presentation, which will be reviewed before the presentation is accepted for the meeting. Some disciplines require presenters to submit a paper of about 6-15 pages, which is peer reviewed by members of the program committee or referees chosen by them.

In some disciplines, such as English and other languages, it is common for presenters to read from a prepared script. In other disciplines such as the sciences, presenters usually base their talk around a visual presentation that displays key figures and research results.

Academic conferences typically fall into three categories:

（1）the themed conference, small conferences organized around a particular topic.

（2）the general conference, a conference with a wider focus, with sessions on a wide variety of topics. These conferences are often organized by regional, national, or international learned societies, and held annually or on some other regular basis.

（3）the professional conference, large conferences not limited to academics but with academically related issues.

Conferences are usually organized either by a scientific society or by a group of researchers with a common interest. Larger meetings may be handled on behalf of the scientific society by a Professional Conference Organiser or PCO.

The meeting is announced by way of a Call For Papers (CFP) or a Call For Abstracts, which is sent to prospective presentersand explains how to submit their abstracts or papers. It describes the broad theme and lists the meeting's topics and formalities such as what kind of abstract (summary) or paper has to be submitted, to whom, and by what deadline. A CFP is usually distributed using a mailing list or on specialized online services. Contributions are usually submitted using an online abstract or paper management service.

参考译文：学术会议

一个学术会议或座谈会是一个为研究者（不一定是学者）召开的介绍和讨论他们工作的会议。会议与学术或科学期刊一起为研究人员之间的信息交流提供了重要渠道。

会议通常由各种演示组成。它们往往简洁扼要，时间跨度约为10至30分钟，通常要在演讲结束后进行讨论。作品可以以书面形式被集合后作为学术论文，并作为会议记录发表。

通常情况下，会议将包括主题发言人（通常是具有某些地位的学者，但有时是来自外部学术界的人士）。主题演讲通常更长，持续时间长达一个半小时，特别是如果一个小组有几个主题发言人。

除了发言之外，会议还举办小组讨论和圆桌会议讨论各种问题，海报会议和讲习班。一些会议还会采用更多的互动模式，例如参与者推崇的"非会议"或各种对话模式。

通常会要求主持人提交简报摘要，并在会议之前进行审查。一些学科要求主持人提交大约6~15页的论文，这是由计划委员会成员或他们所选择的裁判同行评议的。

在一些学科，如英语和其他语言中，演讲者通常从准备好的脚本中读取。在其他学科，如科学，演讲者通常围绕一个可视化界面进行介绍，展示关键数字和研究成果。

学术会议通常分为三类：

（1）主题会议，围绕特定主题组织的小型会议。

（2）大会，更广泛的会议，多个议题的会议。这些会议通常由区域、国家或国际学术社团组织，每年或以其他一些常规方式举办。

（3）专业会议，大型会议不限于学术界，而是与学术有关的问题。

会议通常由科学社团或具有共同兴趣的一组研究人员组织。专业会议组织者或PCO可代表科学社团处理较大的会议。

会议通过"征集论文"（CFP）或"总结摘要"的方式发布，将其发送给主持人，并对如何提交摘要或论文做出解释。它描述的主题很广泛，列出了会议的主题和手续，例如什么样的摘要或论文必须提交给谁以及截止日期。CFP通常使用邮件列表或专门的在线服务进行分发。征稿通常使用在线摘要或纸张管理服务提交。

New Words and Expressions:

academic [ækə'demɪk] *adj.* 学术的；理论的；学院的

panel discussions *n.* 专题讨论会

poster sessions *n.* 海报论文，论文海报展

committee [kə'mɪtɪ]*n.* 委员会

abstract ['æbstrækt]*n.*摘要；抽象；*adj.* 抽象的；深奥的；*vt.* 摘要；提取

prospective [prə'spektɪv] *adj.* 未来的；预期的；*n.* 预期；展望

summary ['sʌməri] *adj.* 简易的；扼要的；*n.*总结

deadline ['dedlaɪn]*n.* 截止期限，最后期限

Questions:

Answer the following questions according to the text.

Write an international conference notice about aircraft design and manufacture which will be held during the winter vacation on your school.

Passage Two: How to Write a Resume—Tips for Writing a Resume for a Job Application

A resume is one of the most important parts of a job application. Your resume is how you tell the story of your professional history to potential employers. Above all, your resume needs to be consistent, concise, and clear and easy to read. If it's not, your resume and cover letter won't get a second glance from any hiring manager.

Read below information on how to write a resume that will get noticed and help you get invited for an interview. How to Write a Resume?

（1）Choose a resume type. There are several basic types of resumes used to apply for job openings. Depending on your personal circumstances, choose a chronological, a functional, combination, or a targeted resume. Taking the time to choose the best type of resume for your situation is well worth the effort.

（2）Choose the right font and size. You want to choose a font and font size that is legible and leaves enough white space on the page. You also want to keep style (such as italics, underlining, bold, and the use of bullets) to a minimum. When you use a particular style, use it consistently.

（3）Review resume examples. Read through samples that fit a variety of employment situations. These sample resumes will provide you with examples of resume formats that will work for almost every type of job seeker. They also help you see what kind of information to include. However, whenever you use a resume example, be sure to customize your resume so it reflects your skills and abilities, and the jobs you are applying for.

（4）Use a resume template. Along with resume examples, you can use a resume template as a starting point for creating your own resume. Add your information to the resume template, then tweak and edit it to personalize your resume, so it highlights your skills and abilities.

（5）Use resume keywords. Most companies use recruiting management software to screen candidates for job openings. In order to get found, your resume needs to contain keywords that directly target the jobs you are interested in. This will also help the hiring manager see how your skills and experiences make you an ideal candidate for the specific job.

（6）Get resume advice. Writing a resume is hard work, and it's often a good idea to get help before you send it to employers. You can find resume writing advice and resume writing tips here. You can also meet with a college career counselor if you are a college student or alumnus. You might use a professional resume service instead, or check with your state's Department of Labor website for information on any free job services they offer. There are many great, free resume resources, so do some research before paying money for someone's advice.

（7）Proof your resume. Be sure to thoroughly edit your resume before sending it. Check for grammar and spelling errors, as well as any style inconsistencies. Consider asking a friend or family member, or even a career counselor, to read over your cover letter. Also review these proofing tips to ensure that your resume is consistent and error free.

参考译文：如何编写简历——求职简历的编写技巧

简历是工作申请中最重要的部分之一。您的简历是如何向潜在雇主讲述关于您的职业史的故事。最重要的是，您的简历需要一致、简洁、清晰易读。否则，您的简历和求职信将不会再从任何招聘经理那里得到第二次阅读。

请阅读以下信息，了解如何撰写简历，使自己了解注意事项并帮助您获得参加面试的邀请。如何撰写简历呢？

（1）选择简历类型。有几种基本类型的简历可用于申请职位空缺。根据您的个人情况，选择时间顺序、功能、组合或有针对性的简历。花时间为您的情况选择最佳类型的简历是非常值得的努力。

（2）选择正确的字体和大小。你想选择一个字体和字体大小是清晰的，并留下足够的空白在页面上，您还希望将风格（如斜体、下划线、粗体和使用项目编号）保持在最低限度。当您使用特定的风格时，一直使用正确的字体和大小。

（3）查看简历示例。阅读适合各种就业情况的样本。这些样本简历将为您提供几乎所有类型求职者工作的简历格式的示例。他们还帮助您了解要包含哪些信息。但是，每当您使用简历示例，请确保自定义简历，以反映您的技能和能力以及您申请的工作。

（4）使用简历模板。除了简历示例，您可以使用简历模板作为创建自己简历的起点。将您的信息添加到简历模板中，然后调整并编辑使您的简历个性化，从而突出您的技能和能力。

（5）使用简历关键字。大多数公司使用招聘管理软件来筛选候选人的职位空缺。为了使您的简历被选中，您的简历需要包含直接针对您感兴趣的工作的关键字，这也有助于招聘经理了解您的技能和经验，使您成为具体工作的理想候选人。

（6）获取简历建议。写简历是艰苦的工作，在将其发送给雇主之前，获取建议帮助通常是一个好主意。您可以在这里找到简历写作建议和简历写作技巧。如果您是大学生或校友，您还可以与大学生涯辅导员会面。您可以使用专业的简历服务，或者向您所在州劳工部网站查询有关其提供的任何免费工作服务的信息。那里有许多伟大的、免费的简历资源，因此在付钱寻求某人建议之前要做一些研究。

（7）检查您的简历。在发送之前，请务必彻底编辑简历。检查语法和拼写错误，以及任何风格的不一致。可以考虑问一个朋友或家人，甚至职业顾问，阅读你的求职信。还要查看这些打样提示，以确保您的简历一致且无错误。

New Words and Expressions:

resume [rɪ'zjuːm] *n*. 摘要；简历；*vt*. 重新开始，继续

concise [kən'saɪs] *adj*. 简明的，简洁的

circumstance [sɜːkəmstəns] *n*. 环境，情况；事件

chronological [ˌkrɑnə'lɑdʒɪkl] *adj*. 按年代顺序排列的

font ['fʌnt] *n*. 字体

legible ['lɛdʒəbl] *adj*. 清晰的；易读的

italics [ɪ'tælɪks] *n.* 斜体字

underlining [ˌʌndɚ'laɪn] *n.* 下划线

bold [bəʊld] *adj.*英勇的；黑体的；险峻的

candidate ['kændɪdeɪt] *n.* 候选人，候补者；应试者

alumnus [ə'lʌmnə] *n.* 男校友；毕业生

Questions:

Answer the following questions according to the text.

Write a resume for a job application according to your major supposing you are graduate this year.

Passage Three: How to Write Academic Papers

1. General Overview

Many students wonder about the writing process itself. The outline of the academic paper is very similar for most branches of science.Creating an extended outline may help structure your thoughts, especially for longer papers. Here are a few samples outlines for research papers.We have also created a couple of articles with general tips and help on research papers.When writing a scientific paper, you will need to adjust to the academic format. The APA writing style is one example of an academic standard frequently used.By the way, here is another great resources on how to write a research paper.

2. Preparing to Write a Research Paper

Usually, the purpose of a research paper is known before writing it. It can be formulated as a research paper question, a thesis statement or a hypothesis statement.If you do not know what to write about, you will have to look for ideas for research paper topics.

3. Structure of a Research Paper

The structure of a research paper might seem quite stiff, but it serves a purpose: It will help find information you are looking for easily and also help structure your thoughts and communication.Here is an example of a research paper. Here's another sample research paper.An empiric paper frequently follows this structure:

（1）Title；

（2）Abstract；

（3）Introduction；

（4）Methodology；

（5）Results；

（6）Discussion；

（7）Conclusion；

（8）References。

4. Additional Parts for Some Academic Papers

The following parts may be acceptable to include in some scientific standards, but may be inappropriate for other standards.Table of Contents (Usually placed right before or right after the abstract). Acknowledgments (Sometimes placed before the abstract and sometimes at the end of a paper). Appendices (Placement: After all the other parts)

Other technical issues:

When correcting papers, we have been surprised by the many students struggling with intext citations. Although the standards are somewhat different, citations in the text should not be too difficult to master. The most used standards for referencing in research papers are APA-standard and MLA-standard.

Some paper formats allow you to include footnotes in the text, while some do not allow footnotes.Authors frequently want to include tables and figures in the text.Sometimes the format or standard prohibits the authors from entering tables and figures directly into the text (where you want your table). Sometimes they have to be included after the main text.

5. Standards and Formats

Please observe that most scientific fields (and paper formats) have their own specific rules and standards of writing.

You will have to check with your faculty or school to know exactly how to write the paper — the guide is meant as an overview of academic papers in general.

6. Publish Articles

Publication of your article can be a very time-consuming process.After writing the academic paper, the researchers submit it to a journal. Typically you start with the most regarded journal and then work yourself down the list, until a journal accepts the article.

参考译文：如何撰写学术论文

1. 概述

许多学生想了解写作过程本身。学术论文的大纲与大部分科学界非常相似。创建扩展轮廓可能有助于构建您的想法，特别是对于较长的论文。以下是研究论文的几个示例轮廓。我们还创建了一些关于研究论文的一般提示和参考文献。撰写科学论文时，您需要根据学术形式进行调整。 APA写作风格是经常使用的学术标准的一个例子。顺便说一下，这里是另一个伟大的资源，关于如何写一篇研究论文。

2. 准备撰写研究论文

通常，研究论文的目的在写作之前是已知的。它可以被制定为研究论文的问题，论文陈述或假设陈述。如果你不知道写什么，你将不得不寻找研究论文的主题。

3. 研究论文结构

研究论文的结构可能看起来很僵硬，但它有一个目的：它将帮助您轻松找到所需的信息，并帮助您构建您的想法和沟通。这是一篇研究论文的例子，是另一份样本研究论文。经验论文经常遵循以下结构：

（1）标题；
（2）摘要；
（3）简介；
（4）方法；

（5）结果；

（6）讨论；

（7）结论；

（8）参考。

4. 部分学术论文的附加部分

以下部分可能被接受以包括在一些科学标准中，但可能不适用于其他标准。目录（通常放置在摘要之前或之后）。致谢（有时放在摘要之前，有时在一篇论文的结尾）。附录（放置：所有其他部分之后）。

其他技术问题：

在纠正论文时，我们感到惊讶的是，许多学生都在试图引用。虽然这些标准有些不同，但文中的引文不应该太难掌握。研究论文中最常用的参考标准是APA标准和MLA标准。

一些纸张格式允许您在文本中包含脚注，而有些则不允许脚注。作者经常希望文本中包括表格和图形。有时，格式或标准禁止作者直接输入表格和图形（您希望在表格中显示）。有时候，它们必须包含在正文之后。

5. 标准和格式

请注意，大多数科学领域（和纸张格式）都有自己的具体规则和写作标准。

您将不得不与您的教师或学校进行确认，以确切地知道如何撰写论文——该指南是为了概括总结学术论文。

6. 发表文章

发表您的文章可能是一个非常耗时的过程。撰写学术论文后，研究人员将其提交给期刊。通常，您从最受关注的期刊开始，然后自己下单，直到杂志接受您的文章。

New Words and Expressions:

outline ['aʊtlaɪn] *n.* 轮廓；大纲；*vt.* 概述描画……轮廓

formulate ['fɔːmjʊleɪt] *vt.* 规划；用公式表示；明确地表达

thesis ['θiːsɪs] *n.* 论文；论点

hypothesis [haɪ'pɒθɪsɪs] *n.* 假设

communication [kə,mjunɪ'keʃən]*n.* 通讯；交流

empiric [em'pɪrɪk] *adj.* 有经验的；成熟的

methodology [meθə'dɒlədʒɪ] *n.* 方法；方法论

reference ['rɛfrəns] *n.* 参考；参考书目

acknowledgment [ək'nɑlɪdʒmənt] *n.* 感谢；承认；承认书

appendices [ə'pendɪsi:z] *n.* 附件；阑尾；附加物

citation [saɪ'teʃən] *n.* 引用；引证

footnotes['fʊt,nəut] *n.* 脚注；附注

journal['dʒɜ:rnl] *n.* 杂志；日记

Questions:

Answer the following questions according to the text.

（1） What is the basic structure of an empiric academic paper？

（2） What would be the recommendation for writting academic papers？

Attachment

Professional Phrase(1)

Pressurization Control 增压控制

Humidity Regulation 湿度调节

Auto Pilot 自动驾驶

Controlling 控制

Indicating 指示

Sensing 传感

Coupling 耦合

Flight Compartment 驾驶舱

Passenger Compartment 座舱

Cargo 货舱

Accessory Compartment 附件舱

Flight Control 飞行操纵

Aileron 副翼

Rudder 方向舵

Elevator 升降舵

Horizontal Stabilizer 水平安定面

Vertical Stabilizer 垂直安定面

Flaps 缝翼

Spoiler & Drag 扰流板与阻力

Airfoil 机翼剖面

Propellers 螺旋桨

Rotors 旋翼

Landing Gear 起落架

Supplementary Gear 辅助起落架

Navigation 导航

Altitude & Direction Inst 高度和航向仪

Radar Navigation 雷达导航

Power Plant 动力装置

Fuselage 机身

Cones & Fillets/Fairings 尾锥和整流片

Propeller Assembly 螺旋桨组装

Professional Phrase(2)

Aerodynamic coefficients: nondimensional coefficients for aerodynamic forces and moment.

空气动力系数：空气动力和力矩的无量纲系数。

Aircraft arresting device: an apparatus at least part of which is external to the aircraft，designed to transmit energy to an arresting mechanism to arrest or stop an aircraft on the landing oraborted takeoff area.

航空器阻拦装置：其一部分至少是在航空器外部的一种器具，设计成将能量传送到阻拦机构上，使航空器被阻拦或停止在着陆区或中断起飞区。

Aircraft configuration: a term referring to the relative position of the various elements affecting the aerodynamic characteristics of an aircraft.

航空器构形：叙述影响航空器气动特性的各种部件相对位置的一个术语。

Aircraft engine: means an engine that is used or intended to be used for propelling aircraft.

航空器发动机：指用于或拟用于推进航空器的发动机。

Amphibian :an aircraft designed to take off and land on either land or water.

水陆两用航空器：设计成可在陆地或水面起降的一种航空器。

Angle of bank :the angle between the lateral axis of an aircraft and the horizontal. Also calledangle of roll.

倾斜角：航空器横轴与水平面的夹角。

Angle of descent: the angle between a decending aircraft's flight path and the horizontal.

下降角：正在下降的航空器的飞行航迹与水平面的夹角。

Angle of glide : the vertical angle between the flight path of a gliding aircraft and the horizontal.

滑翔角：滑翔飞机的飞行路径与水平之间的角度。

Automatic tracking: a device which automatically follows a selected target using radar inputsignals and provides continuous positional data in analog or digital form.

自动跟踪：利用雷达输入信号自动跟踪选定目标，并以模拟或数字方式提供连续位置数据的装置。

Autopilot: those units and components which furnish a means of automatically controlling the aircraft.

自动驾驶仪：对航空器提供自动控制手段的装置和组件。

Autopilot couple: the means used to link an autopilot to a navigation system for Automatic flight control.

自动驾驶仪耦合器：为实现自动飞行控制目的而将自动驾驶仪与导航系统相连所使用的装置。

Auxiliary rotor: a rotor that serves either to counteract the effect of the main rotor torque on a rotorcraft or to maneuver the rotorcraft about one or more of its three principal axes.

辅助旋翼：用于平衡旋翼机主旋翼扭矩或使旋翼机绕其三根主轴线中一根或一根以上作机动飞机的旋翼。

Axis of no feathering:in rotorcraft, the axis about which there is no first harmonic feathering orcyclic pitch variation of the rotor system.

无顺桨轴线：旋翼机的一根轴线，绕此轴线没有旋翼系统的一次协调顺桨或周期变距。

Blade angle: the angle that a blade makes with some reference plane, e.g. the angle between the geometric cord or zero-lift cord of a propeller blade and a plane perpendicular to the axis of rotation of the propeller. In rotor craft, the acute angle between the blade cord line and the plane of rotation measured at a predetermined reference station of the blade.

桨叶角：桨叶与某基准面之间形成的角，例如螺旋桨桨叶几何弦或零升力弦与垂直于螺旋桨转轴的平面之间的夹角。在旋翼机上，桨叶角指在旋翼预定基准站位处测得的桨叶弦线与旋转面之间的锐角。

Blade damper: in rotor craft，a device, such as a hydraulic damper, installed to act about the draghinge of a rotor blade to reduce horizontal oscillation.

桨叶减摆器：旋翼机上一种装置，例如液压减摆器，它被安装作为旋翼桨叶的垂直铰，以减小水平摆动。

Blast fence: a barrier that is used to divert or dissipate jet or propeller blast.

导流栅：用以使喷气或螺旋桨气流转向或扩散的阻挡装置。

Booster:a device that provides extra power,thrust,or pressure.

助力器：提供额外功率、推力或压力的装置。

Breakhorse power: the power delivered at the propeller shaft (main drive or main output) of an aircraft engine.

制动马力：航空器发动机螺旋桨轴(主传动轴或主输出轴）发出的功率。

Buffeting:the condition in which repeated aerodynamic forces are experienced by any part of anaircraft which have been caused and maintained by unsteady flow arising from a disturbance setup by any other part of the aircraft.

抖振：航空器任一部分经受反复气动力的一种状态，它是由航空器任一其他部分产生的扰动造成的非定常气流引起并维持的。

Cartwheel: a maneuver or stunt in which an airplane is snap rolled from a steep bank in one direction to a steep bank in the other direction, reversing its original direction of flight.

半滚跃升倒转：一种机动飞行或特技飞行，此时飞机从一个方向的大坡度快滚到另一方向的大坡度，飞行方向改变180度。

Catapult:a device that applies external power to an airplane to launch it at flying speed without the use of a conventional runway.

弹射器：向飞机施加外部动力，将其以飞行速度发射而不需使用常规跑道的一种装置。

Cockpit navigation:navigation performed by pilots at their normal duty station that is capable of meeting domestic or oceanic navigation requirements.

驾驶舱导航：驾驶员在其正常工作位置上完成的导航，能符合国内或远洋导航要求。

Collision avoidance system: an airborne system that performs all the necessary functions such that its output is a signal indicating or initiating an appropriate avoidance maneuver at asuitable time.

防撞系统：一种机载系统，它执行所有必要的功能，以使其输出的是一种在适当时刻指示或开始采取相应规避机动动作的信号。

Cowl flap: one of several shutters in an aircraft engine cowing, used to regulate the flow of cooling air around the engine.

整流罩风门片：航空器发动机整流罩中若干节气之一种，用以调节发动机周用冷却空气的流量。

Critical engine:the engine whose failure would most adversely affect the performance or handling qualities of an aircraft.

临界发动机：其失效将最严重地影响航空器性能及操纵品质的那台发动机。

Dive brake:a flap or removable surface which, when extended, reduces the speed of the aircraft in a dive.

俯冲减速板：打开时可减小俯冲中航空器速度的襟翼或活动面。

Hood: a collapsible hood or opaque shield used in instrument flight instruction to prevent the person practicing instrument flight from seeing outside the cabin or cockpit.

暗舱罩：仪表飞行训练中使用的可收放的罩或不透光的遮盖物，防止进行飞行实习的人员看到座舱或驾驶舱外面的景物。

Hydroplaning: the condition in which moving aircraft tires are separated from a pavement surface by a water or liquid rubber film, or by steam resulting in a derogation of mechanical braking effectiveness.

滑水现象：运动中的航空器轮胎因水或液体磨擦膜，或因水雾致机械刹车效率减少而脱离跑道面的一种现象。

Jet assisted takeoff (JATO): the use of supplemental jet or rocket engines to assist an aircraft totakeoff in less than the normally required distance or in the event of failure of the primary power source.

暇气助推起飞（JATO）：利用辅助的喷气或火箭发动机协助航空器在少于正常需要的距离上或在主动力源万一失效的情况下起飞。

Jet blast: high energy wind forces accompanied by critical air movement velocities created by the Exhaust of turbojet engines.

喷流冲击波：涡轮喷气发动机排气造成的具有临界空气运动速率的高能量风力。

Leading edge: the edge of an airfoil or propeller blade which first meets the air.

前缘：翼面或螺旋桨叶片首先遭遇空气的边缘。

Mainbang: transmitted pulse, within a radar system.

主脉冲：雷达系统内的发射脉冲。

Mainrotor: the rotor that supplies the principal lift to a rotorcraft.

主旋翼：提供旋翼机主要升力的旋翼。

Maneuver:as pertaining to aircraft, an intended change in the movement or attitude of an aircraft in flight, such as a turn, a dive, a pullout, a bank, etc. An evolution or performance such as aloop, a roll, a wing-over, etc.

机动飞行：用于航空器时，指有意地改变飞行中航空器的运动或姿态，例如转弯、俯冲、改出、倾斜斤斗、横滚、跃升转弯半倒转等的变化或性能。

Mean aerodynamic chord: the cord of an imaginary rectangular airfoil that would have pitching moments throughout the flight range the same as those of an actual airfoil or combination of airfoils under consideration, calculated to make equations of aerodynamic forces applicable.

平均气动弦：假想矩形机翼的翼强，此机翼在整个飞行范围内的俯仰力矩与所考虑的真实机翼或机翼组合的俯仰力矩相同,平均气动弦的计算是为了使气动力方程更实用。

Minimum obstruction clearance altitude (MOCA): that specified altitude in effect between radio fixes on VOR airways,off- airway routes, or route segments which meets obstruction clearance requirements for the entire route segment and which assumes acceptable navigational signal coverage only within 22 nautical miles of a VOR.

最小越障高度（MOCA）： VOR 航路、偏离航路航线或航线段上两个无线电定位点之间的规定高度，它在整个航段上均符合越障要求，并且仅在VOR 的22 海里范围内具有可接受的导航信号覆盖能力。

Minimum reception altitude (MRA): the lowest altitude required to receive adequate signals to determine specific VOR/OVRTAC/TACAN fixes.

最小接收高度（MRA）：为接收足够的信号以确定具体的VOR/OVRTAC/TACAN 位置所要求的最低高度。

Minimum sector altitude: the minimum IFR altitude in a quadrant within 25 nautical miles of a navigational aid upon which an instrument approach procedure is predicated.

最小扇形区高度：导航辅助设施25 海里范围内某一扇形信号区的最小仪表飞行规定高度，按此高度预定好仪表进近程序。

Minimum vectoring altitude (MVA): the lowest altitude expressed in feet above mean sea level,that an instrument fight rules aircraft will be vectored by a radar controller. That altitude assures communications, radarcoverage and meets obstacle clearance criteria.

最小引导高度（MVA）：由雷达操纵员引导仪表飞行规则的航空器飞行的最低高度，用平均海平面以上的英尺数表示。此高度保证通信、雷达覆盖并符合越障要求。

Oilburner: a military operation simulating jet aircraft bombing runs.

燃烧器:模拟喷气式航空器轰炸飞行的军事行动的设备。

On course indication (aural): the continuous 1,020 Hz tone heard by the pilot while flying at ornear the center of a radio range course.

在航向指示（音响式）：在无线指向标波束中心上或中心附近飞行时，驾驶员听到的连续1 020 赫音调。

On course indication (visual): the loci of points at which a zero reading on the micro ammeter, cross-pointer, or pilot' s direction indicator is received.

在航向指示（目视式）：接收到微安表、交叉指针式指示器或航向指示器上零读数的点的轨迹。

On course signal: a signal indicating that the aircraft in which it is received is on course，following a guiding radio beam.

在航向信号：一种信号，表示接收到此信号的航空器正处在航向上，跟踪着导引无线电波束。

Pants: a set of teardrop-shaped fairings around the wheels of a fixed landing gear on certain airplanes.Also called spats.

机轮整流罩：某些飞机上围绕固定式起落架机轮的一套水滴形整流罩。又称轮罩。

Pitch: ① the blade angle of a propeller or rotor blade; ②the movement of an aircraft about it slateral axis.

桨距；俯仰：①螺旋桨或旋翼桨叶的桨叶角；②航空器绕其横轴的运动。

Pitchsetting: the propeller blade setting as determined by the blade angle measured in a manner，and at a radius, specified by the instruction manual for the propeller.

桨距调定：螺旋桨桨叶调定，它由按照螺旋桨说明书规定的方法，在其规定的半径上测得的桨叶角来决定。

Plane of rotation: the plane in which an object such as a propeller rotates, specifically, a plane in which a rotary-wing system rotates, or is assumed to rotate.

旋转平面：物体（如螺旋桨）旋转形成的平面，特别是旋翼系统旋转或假定旋转形成平面。

Propeller: a device for propelling an aircraft that has blades on an engine-driven shaft and that，when rotated, produces by its action on the air, a thrust approximately perpendicular to its plane of rotation. It includes control components normally supplied by its manufacturer, but does notinclude main and auxiliary rotors or rotating airfoils of engines.

螺旋桨：推进航空器的一种装置，在其由发动机驱动的轴上装有桨叶，转动时靠其对空气的作用产生与其旋转面几乎垂直的拉力。它包括通常由其制造商供应的操纵部件，但不包括主旋翼和辅助旋翼或发动机的旋转叶片。

参考文献

[1] 刘星. 飞行原理[M]. 北京：科学出版社，2011.

[2] Ramesh K. A. Recent Progress in Some Aircraft Technologies [M]. Croatia: lTexLi，2016.

[3] 刘继新. 飞行专业英语阅读[M]. 北京：国防工业出版社，2014.

[4] 胡静. 航空器制造工程专业英语[M]. 西安：西安电子科技大学出版社，2015.

[5] 蔡忠元，吴鼎民，昂海松. 航空专业英语教程[M]. 北京：航空工业出版社，2008.

[6] Richard L. B. Rotary Wing Structural Dynamics and Aeroelasticity [M]. 北京：航空工业出版社，2014.

[7] 刘志武. 航空专业英语[M]. 北京：北京理工大学出版社，2015.